Component-Based Software Engineering

Component-Based Software Engineering

Methods and Metrics

Umesh Kumar Tiwari and Santosh Kumar

CRC Press
Taylor & Francis Group
Boca Raton London New York

CRC Press is an imprint of the
Taylor & Francis Group, an **informa** business

A CHAPMAN & HALL BOOK

First edition published 2021
by CRC Press
6000 Broken Sound Parkway NW, Suite 300, Boca Raton, FL 33487-2742

and by CRC Press
2 Park Square, Milton Park, Abingdon, Oxon, OX14 4RN

Library of Congress Cataloging-in-Publication Data

ISBN: 978-0-367-35488-6 (hbk)
ISBN: 978-0-429-33174-9 (ebk)

Typeset in Palatino
by SPi Global, India

To the Almighty and Our Families

Contents

Figures and Tables

Figures

Tables

Preface

In the current software development environment, the "divide and conquer" approach is used in development of very large-scale and complex software. Software is divided into components and then individual components are either developed or purchased by the development team or by the third party. Finally, developed/purchased components are integrated to design the software according to the requirements. This development approach is known as component-based software engineering (CBSE). Component-based software engineering, a distinctive software engineering paradigm, offers the feature of reusability. It promotes the development of software systems by picking appropriate pre-existing, pre-built, pre-tested and reusable software work products called *components*. These components are assembled and integrated within a well-defined architectural design. Rather than focusing solely on coding, component-based software development enables application developers to concentrate on better design and optimized solutions to problems, since coding objects are available in the repository in the form of components. Component-based software engineering accentuates development *with* reuse as well as development *for* reuse. Reusable components interact with each other to provide and access functionalities and services. The interactions and integrations of heterogeneous components raise issues, including the suitable and efficient reusability of components, complexities produced during the interaction among components, testing of component-based software and the overall reliability of the application under development.

This book addresses four core issues in the context of component-based software: *reusability*, *interaction/integration of components*, *testing* and *reliability*. In addition, the major research issues addressed in this book are:

- Lack of selection techniques for components from the off-the-shelf repository
- Lack of maintenance techniques for components from the off-the-shelf repository
- Lack of empirical and experimental software describing the implementation of theoretical architectures
- Conflict between usability and reusability
- Lack of integration and adaptation methods for components at runtime
- Lack of empirical and experimental metrics
- Measurement methods for the integration complexities of components
- Measurement and methods for the interaction complexities of components
- Lack of certification criteria and procedures for selection/development/outsourcing of individual components
- Lack of verification criteria and procedures for selection/development/outsourcing of individual components
- Lack of certification criteria and procedures for integrated CBSE software
- Lack of verification criteria and procedures for integrated CBSE software
- Lack of testing levels for individual components and integrated CBSE software

- Lack of testing procedures and methodologies for individual components and integrated CBSE software
- Reliability and sensitivity to changes.

In this book, these issues are addressed using a model-driven approach to analyze and show the results of proposed measures and metrics. This approach is a very simple, suitable and comparatively efficient way to resolve them effectively. Appropriate scenarios and case studies are constructed to model the component-based environment. Comparative results are reported in each chapter, demonstrating the efficiency of the proposed metrics.

Intended Audience

This book focuses on a specialized branch of the vast domain of software engineering: component-based software engineering. It is designed for Ph.D. research scholars, M. Tech. (computer science and engineering) scholars, M. Tech. (information technology) scholars, B. Tech. (computer science and engineering) students, B. Tech. (information technology) students and other higher education courses, as well as for research areas where software engineering is in the syllabus.

Key Benefits of this Book

This work not only reviews and analyzes previous work by eminent researchers, academics and organizations in the context of CBSE, but also suggests some innovative, efficient and better solutions.

1. This book provides a proper mechanism to assess the reusability of individual components.
2. This book provides metrics and measures to assess the reusability of component-based software.
3. This book addresses the problem of component selection from the repository.
4. In this book we have included the complexity metrics with respect to assessing the integration and interaction of components.
5. Testing techniques for CBSE, black box and white box are covered.
6. We have provided testing techniques to assess the number of test cases in both contexts: black box and white box.
7. This book covers reliability estimation techniques for CBSE.
8. Major issues in the context of component-based software are discussed in detail.
9. A detailed literature survey covers reusability, complexity, testing, reliability and quality issues.

Chapter Organization

This book contains seven chapters, organized as follows:

Chapter 1 introduces the basic concepts of software engineering in terms of traditional development and advanced development paradigms. Further, it defines the context of component-based software engineering (CBSE), including definitions of components and CBSE, evolution of the CBSE paradigm, basic characteristics of CBSE, related problems and observations.

Chapter 2 provides comparative review and analysis of various component-based development models proposed and used so far. These models include the BRIDGE, Umbrella, X, Y and other models.

Chapter 3 explores the major issues in developing component-based software, focusing on reusability, interaction/integration complexities, design and coding, testing and reliability. The chapter provides a detailed comparative literature review of these issues.

Chapter 4 focuses on the reusability features of components and the role of reusability in selection and verification of components. In this chapter metrics for reusability are developed at component level as well as at system level. These metrics are categorized according to the degree of reusability as new, adaptable or off-the-shelf components.

Chapter 5 defines some simple and effective interaction and integration metrics to assess the complexities of component-based software. These interaction metrics are divided into two categories: black-box and white-box. For black-box components, integration metrics are developed and for white-box components, cyclomatic complexity metrics are constructed.

Chapter 6 focuses on testing and test-case generation issues in component-based software. Two types of technique are defined in this chapter: one for components whose code is not accessible and the other for those whose code is available. In the context of CBSE, testing and the assessment of reliability is one of the crucial issues.

Chapter 7 illustrates some measures to assess the execution time and reliability of component-based applications using reusability feature of the components. To enhance the reliability of component-based software, interaction metrics and reusability metrics are used as the reliability estimation factor.

Acknowledgements

No creation in this world is a solo effort, and neither is this work for which we are credited. From the person who inspired us to write this book to the person who approved the work, everyone has a role. We have benefited from various resource materials and ideas on component-based software engineering, and we are gratified that over the entire period of the compilation of this book, many people have helped and supported us.

It is our pleasure to express our gratitude to Prof. (Dr.) Kamal Kumar Ghanshala, Hon. President, Graphic Era (Deemed to be University), Dehradun for encouraging us and providing the resources and facilities to conduct this work. We express our profound gratitude to Prof. (Dr.) R. C. Joshi, Chancellor, Graphic Era (Deemed to be University), Dehradun for his wise suggestions, never-ending inspiration and deep interest in this book. We express our thanks to Prof. (Dr.) Rakesh K. Sharma, Vice Chancellor, Graphic Era (Deemed to be University), Dehradun and Prof. (Dr.) H. N. Nagaraja, Pro-Vice Chancellor, Graphic Era (Deemed to be University), Dehradun for their guidance and support.

We are very thankful to Dr. D. R. Gangodkar, Dr. Pravin P. Patil, Dr. Bhasker Pant, Dr. Devesh Pratap Singh and Mr. Manish Mahajan for their valuable advice and continuous evaluation of this book. We are especially grateful to Prof. Dibyahash Bordoloi, Prof. (Dr.) S.C. Dimri, Dr. Priya Matta and Dr. Kamlesh Purohit for their academic, administrative and moral support. We thank all our dear colleagues who helped us unconditionally and selflessly during the writing of this work.

We would like to thank students, readers of this book and practitioners, and hope that this work will be helpful for them in their research in the field of component-based software engineering. We would also like to thank Ms. Balambigai, Senior Project Manager, SPi Global, Ms. Aastha Sharma and Ms. Shikha Garg at CRC Press/Taylor & Francis Group for their continuous support and cooperation throughout the publication process.

We express our deep sense of appreciation to our families for their unconditional love, support and care.

Umesh Kumar Tiwari
Santosh Kumar

Authors

Umesh Kumar Tiwari is an associate professor in the Department of Computer Science and Engineering in Graphic Era (Deemed to be University), Dehradun, India. He received his Ph.D. from Graphic Era (Deemed to be University) in 2016 and Master of Computer Applications in 2005. He has more than 13 years of research and teaching experience in AICTE/UGC-approved universities and colleges at undergraduate and postgraduate levels. He is the author of two books, *Principles of Programming Languages* (A.B. Publication, Delhi, 2012) and *Paradigms of Programming Languages* (A.B. Publication, Delhi, 2010). He supervises Ph.D. and M. Tech students for their theses. He has published more than 35 research papers in national and international journals/conferences in the field of software engineering, wireless networks, WSN and IoT. His research interests include component-based software engineering, wireless networks, WSN, IoT and Agile methodology.

Santosh Kumar received his Ph.D. from IIT Roorkee (India) in 2012, M. Tech. (CSE) from Aligarh Muslim University, Aligarh (India) in 2007 and B.E. (IT) from C.C.S. University, Meerut (India) in 2003. He has more than 13 years of experience researching and teaching undergraduate (B. Tech.) and postgraduate (M. Tech.) courses as a lecturer, assistant professor and associate professor in various academic/research organizations. He has supervised 01 Ph.D. thesis, 25 M. Tech. theses and 18 B. Tech projects, and is currently mentoring six Ph.D. students (singly and jointly) and eight B. Tech. students. He completed a consultancy project entitled "MANET Architecture Design for Tactical Radios" for DRDO, Dehradun from 2009 to 2011. He is an active review board member for various national and international journals—*IEEE Transactions on Computational Social Systems, IEEE Access, ACM Transactions on Cyber-physical Systems, Soft Computing* (Springer)—and conferences. He is a senior member of ACM and a member of IEEE, IAENG, ACEEE and ISOC (USA), and has contributed more than 65 research papers to national and international journals/conferences in the field of software engineering, wireless networks, WSN, IoT and machine learning. He is currently professor at the Graphic Era (Deemed to be University), Dehradun, India. His research interests include software engineering, wireless networks, WSN, IoT and machine learning.

1

Introduction to Software Engineering and Component-Based Software Engineering

1.1 Introduction

In last few years, the nature of software as well as the software development process has changed significantly. Development processes have come a long way from ad-hoc development to a customer-specific agile development process, from simple, single-task programs to complex and extremely large component-based software, from mere number-crunching mathematical calculations to real-life, problem-specific and dynamic solutions. In today's era, software is not merely a product: it's a medium for delivering products, it's a service.

In computer science, two fundamental constructs of developing a program or software are algorithms and data (structure). Generally, algorithms and logics are developed to create or manipulate data. These algorithms are non-executable codes that are translated into executable form, commonly known as programs. Further, these executable programs are collectively grouped together to form the executable software.

In the modern era, however, software is more than a collection of executable programs. Today's software is a combination of executable programs that perform defined tasks, data and data structures on which these programs will operate, operating procedures, and associated documentation to help users and customers understand the software (see Figure 1.1).

1.2 Software Engineering

Software engineering is both an art and a science. It is the process of constructing acceptable artifacts with scientific verifications and validations within the limitations of time and budget. The term "software engineering" came into existence in the 1960s when NATO's Science Committee organized conferences to address the "software crisis." Krueger (1992) describes the software crisis as "the problem of building large, reliable software systems in a reliable, cost-effective way." Previously, the industry, the research community and academia had concentrated on the development of capable and competent hardware. The result of this effort was the availability of powerful and cheaper machines. Now, there was the requirement of large, functionally efficient software to fully utilize the capability of the available hardware machines and other resources. Thus, the focus of the community

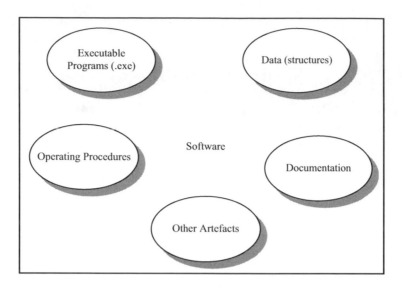

FIGURE 1.1
Modern software.

shifted from small, number-crunching programs to developing software that could address real-world problems. The outcome of these conferences was the emergence of the term "software engineering."

Software is a set of executable programs, related documentation, operational manuals and data structures. Engineering is the step-by-step evolution of constructs in an organized, well-defined and disciplined manner. Some leading definitions of software engineering produced by eminent researchers and practitioners are as follows:

> Software Engineering is the establishment and use of sound engineering principles in order to obtain economically developed software that is reliable and works efficiently on real machines.
>
> **(Bauer et al. 1968)**
>
> Software Engineering is the application of science and mathematics by which the capabilities of computer equipment are made useful to man via computer programs, procedures, and associated documentation.
>
> **(Boehm 1981)**
>
> Software Engineering is the systematic approach to the development, operation, maintenance, and retirement of software.
>
> **(IEEE 1987)**
>
> Software Engineering is a discipline whose aim is the production of quality software, software that is delivered on time, within budget, and that satisfies its requirements.
>
> **(Schach 1990)**
>
> The application of a systematic, disciplined, quantifiable approach to the development, operation, and maintenance of software; that is, the application of engineering to software.
>
> **(IEEE Standard 610.12-1990 1993)**

The conclusion of all these definitions is that the ultimate goal of software engineering is to develop quality software within reasonable time limits and at affordable cost. Software engineering is an engineering domain that seeks to implement standard ways of developing and maintaining software, through standard methods using standard tools and techniques.

1.3 Paradigms of Software Engineering

Since the inception of software, many software development paradigms have evolved, flourished and been well received by academia, the research community and the industry. These paradigms act as a guiding tool for the development community, not only making the development process systematic and disciplined, but also making the life cycle of their developed product easy for users to understand. Here we present a systematic analysis of these paradigms. On the basis of their evolution, we categorize all development models into two broad categories: traditional software engineering paradigms and advanced software engineering paradigms.

1.4 Traditional Software Engineering Paradigms

With the evolution of the term "software engineering," a number of software development paradigms came into existence. These traditional software engineering models can be categorized into three broad categories: classic life-cycle paradigms, incremental paradigms and evolutionary paradigms. All these paradigms have a finite set of development steps starting with requirement gathering and progressing through to deployment. Traditional software engineering paradigms and their categories are shown in Figure 1.2.

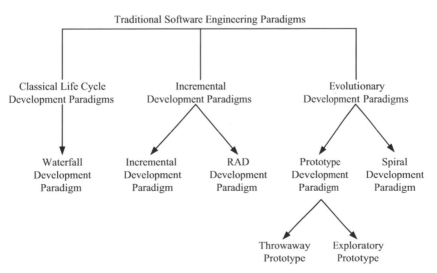

FIGURE 1.2
Traditional software engineering paradigms.

1.4.1 Classic Life-Cycle Paradigms

These are the fundamental development paradigms that came into existence just after the first NATO conference on software engineering in 1967. The development phases of such paradigms are based on the development life cycle of systems previously available in the literature.

1.4.1.1 The Basic Waterfall Development Paradigm

One of the first models to document development as a defined and disciplined process was the basic waterfall model. It was linear in nature, with the accomplishment of one phase followed by commencement of the next. As one of the first development paradigms, basic waterfall was simply an organized, step-by-step engineering software process. Waterfall does not take much account of changes in requirements, adaptation of new enhancements during development, or regular feedback from customers.

The waterfall model follows an orderly chronological approach to developing software which commences with a requirements analysis and progresses through design, implementation, testing, deployment, and finally, maintenance, as shown in Figure 1.3. Winston Royce (1970) did not regard the waterfall model as linear. He made provision for backward loops or feedback loops for every phase of development.

1.4.1.1.1 Key Findings

Key features of this paradigm are

- One of the first paradigms based on the software development life cycle, having a defined set of software development phases.
- The waterfall paradigm freezes all customer requirements at a very early stage, so the selection of the development team, and acquisition of software and hardware resources is done before the actual development begins.
- Waterfall is a very simple paradigm and comparatively easy to implement.

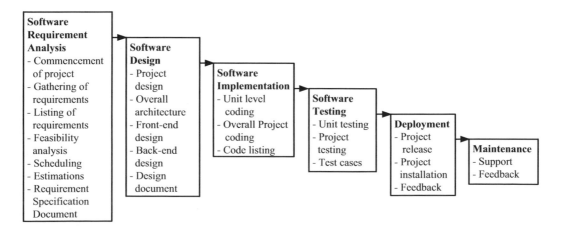

FIGURE 1.3
Basic waterfall development paradigm.

- Development phases are organized in such a way that they do not overlap, that is, the new phase commences only after the conclusion of the previous phase.
- Overall development takes less time as this paradigm requires all requirements to be listed in a clear and predefined manner. Once all the requirements are documented, developers can focus solely on core development.
- The development process is manageable and always under control, as the team's full emphasis is on one phase at a time.
- Each phase has a defined and definite deliverable. After the accomplishment of each phase the work product can easily be reviewed.

This paradigm is best suited to projects where all the requirements are clear and defined, and the development team is sure about the tools and technologies required for the development. The waterfall model is suitable where reengineering of old software is required, and is performed with the help of new technologies and on new platforms. This model is used when the development team has enough experience to develop similar nature of software.

1.4.1.1.2 Critiques

- It is rare for all software requirements to be available at the very early stages of development. Sometimes, the customer is not actually aware of or clear about the requirements, in which case the development process cannot proceed.
- Customers or users are not involved during the development process of their own software other than in the first phase, so they are required to be patient till the last phase, which is quite difficult.
- It is a rigid paradigm which does not allow for flexibility, or changing or enhancing requirements in between the development phases.
- The sequential nature of the waterfall paradigm produces "blocking states" (Bradac et al. 1994), that is, the development team cannot start the next phase until they have finished the current phase. Parallel development phases are not described in the waterfall paradigm. This sequential approach results in coders and testers remaining idle for a long time.
- Errors made by systems analysts, designers or coders are not detected till the final phase of development, as the testing phase comes almost at the end of the paradigm. A high level of uncertainty and risk is therefore involved in the waterfall paradigm.
- Wastage of resources is quite possible as all the resources are acquired long before the commencement of actual development.
- This paradigm is not suitable for innovative projects where the requirements are not very clear, or uncertainty or risk in the project is likely to occur.

1.4.2 Incremental Development Paradigms

Incremental paradigms sought to overcome some of the disadvantages of the basic waterfall model. In these paradigms, development phases are sequential, like waterfall, but they do not require the freezing of all requirements in the first phase. They appreciate working software in a short period of time, but with limited functionality. They deliver software in increments. Development can be started with a small number of requirements, with other requirements added in the next increment.

Incremental development paradigms can be categorized into two classes:

(a) The incremental paradigm

(b) Rapid application development

1.4.2.1 The Incremental Development Paradigm

The incremental paradigm implements the linear and sequential approach of waterfall development in an iterative manner. Each iteration includes a small set of requirements and produces a working version of the software. The incremental paradigm can start development with the set of requirements that is initially available. The deliverable of the first increment is the "core product." In the next increment, additional customer requirements can be added. Each increment contains additional features based on the customer's extended requirements (Figure 1.4).

1.4.2.1.1 Key Findings

- Huge and complex software may be componentized or modularized and necessary modules can be delivered with available resources.

- Development can be started with an initial or most prioritized set of requirements, without the need to freeze all the requirements in advance.

- The customer obtains operational software with basic functionalities within a short time.

- On the basis of customer feedback, the subsequent increments can be better planned, and developers have the time to produce better-quality deliverables that satisfy the customer.

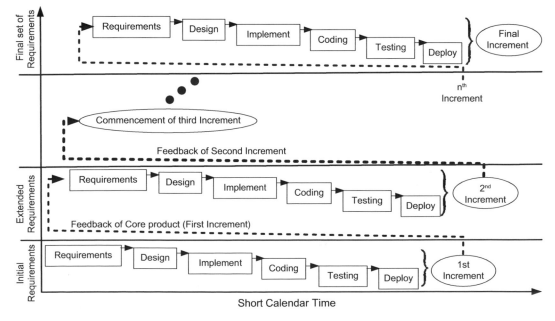

FIGURE 1.4

The incremental development paradigm. (Adapted from Pressman 2005.)

- Incremental development can be started with a limited development team, as the requirements are comparatively less.

- Errors made in the core product or first increment can be easily rectified in the next iteration. Since the size of each delivered product is low, testing and debugging is easy and can be achieved within justifiable costs and time.

- The incremental paradigm is more flexible than the waterfall paradigm in terms of changes in customer requirements. Developers can incorporate extended or changed requirements in subsequent increments.

1.4.2.1.2 Critiques

- Increments should be planned and designed so that each increment is organized in a defined manner. Each increment should add functionality rather than cost overheads and development time.

- The incremental paradigm may be started with the highest-priority requirements, but prioritization of requirements is a complex task. The core product may become irrelevant if the customer's priorities have changed in the next increment.

- Cost and time to incorporate new functionalities in the working software may be higher than in the waterfall paradigm.

- The integration of new or changed requirements in the operation software may lead to new and unforeseen errors, which may cause the budget and development schedule to overrun.

1.4.2.2 The Rapid Application Development (RAD) Paradigm

RAD is one of the first paradigms to appreciate not only the concept of componentization of complex software, but also the parallel development of components. The RAD paradigm follows the "divide and conquer" approach, in which large projects are divided into smaller modules which are allotted to different development teams. Teams develop modules in parallel, and finally these modules are integrated and tested. The division of modules is based on the requirements and early-stage planning by analysts and systems engineers. Every development team works in coordination with other teams as well as with the requirements elicitation and planning teams. The scope and nature of the RAD paradigm differ from those of other paradigms, as it focuses not only on development but also on the better utilization of the development team as well as the resources available.

Componentization or modularization of software starts after the requirements and overall analysis phases. One of the major goals of the RAD paradigm is to shorten the overall development schedule and deliver the product in the shortest possible time. The RAD architecture is shown in Figure 1.5.

1.4.2.2.1 Key Findings

- The RAD paradigm includes advanced modeling tools and techniques. Its design phase consists of three modeling concepts: business modeling or logic building, data modeling and process modeling.

- RAD supports auto code generation mechanisms and code reusability.

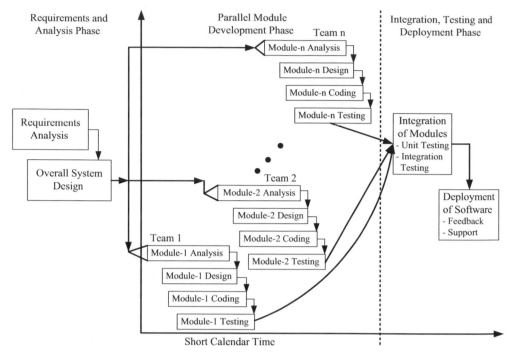

FIGURE 1.5
The rapid application development paradigm. (Adapted from Pressman 2005.)

- RAD is one of the first paradigms to apply the partitioning of large problems into smaller sub-problems. RAD follows the "divide and conquer" approach to development.

- Componentization and modularization are among the fundamental attributes of the RAD paradigm. Large and complex software is divided into modules and these modules are allotted to different development teams.

- Parallel development of different modules results in a shorter life cycle for the overall software. Basically, RAD is used when the customer needs large-scale software quickly. In general, RAD is a "high-speed" version of classic life-cycle paradigms.

- It is the most suitable model for business-oriented products where market-bounded time constraints govern product launches. Some authors have therefore used the terms "business modeling" and "data modeling" in place of analysis and design.

- RAD offers enough flexibility to accommodate changes in or extensions to the customer's requirements. As it is an incremental paradigm, changes can be incorporated and delivered in increments.

1.4.2.2.2 Critiques

- RAD uses extensive human resources, as it requires various teams to develop different modules.

- Integration of different modules developed by different teams with varying levels of technical skill is a major issue. If not handled properly, integration and interaction of modules may generate unmanageable complexities.

- When the assembled modules are tested as a complete system, integration testing and regression testing may be time-consuming as well as costly.

- Coordination among development teams is a critical issue. Different teams have different technological expertise. Combining their work on a common platform and utilizing their skills to achieve a common goal (customer requirements) may present a real challenge.

- RAD requires experienced and skilled systems analysts and designers. Modularization of software is a risky and complex job. If division of modules is not done properly, or not according to the requirements, it will create design faults, which may ultimately result in faulty software.

- The initial cost of the RAD paradigm is comparatively high, as it requires various development teams, and each team requires resources.

1.4.3 Evolutionary Development Paradigms

With the passage of time, customer requirements either change or extend. With the evolution of new technologies and new platforms, software takes on new shapes and dimensions. In such a dynamic and rapidly changing environment, linear development cannot satisfy the needs of customers unwilling to wait until the final phase of development. Evolutionary paradigms are based on the approach that "software evolves over time" (Gilb 1988). Large and complex software is developed in iterations. Customers get the flexibility of being involved in each iteration.

Evolutionary paradigms are classified under two broad paradigms:

1. Prototyping development paradigm
2. Spiral development paradigm

1.4.3.1 The Prototyping Development Paradigm

Prototyping is the most suitable paradigm for new and innovative projects where the customer is unsure of their requirements and the development team is unsure about the technology being used to fulfill these requirements. Prototyping can be scaled up to any level, that is, it is suitable for small-scale innovative projects as well as for large and complex software, where inputs, outputs, logic, technology or the behavior of the software is not clearly identifiable at the commencement of the project.

Moreover, prototyping is actually not a development paradigm, but a way of refining and identifying the actual requirements of the system to be developed. After refining the customer's needs and the requirements of the proposed software, prototyping can be associated with any development paradigm discussed in this section. Figure 1.6 shows the basic working of the prototype paradigm.

The prototyping paradigm starts with interactive meetings between customers and development teams. Both parties try to understand the overall behavior and intentions of the software. They focus on configuring refined and more concentrated requirements from what is available, trying to define undefined problems, and attempting to identify the platforms as well as the technologies that can be used to design and develop the software. On the basis of these findings, a rapid analysis and design is performed. The developers then make prototype systems which are further evaluated by customers. Based on the customer feedback, the next iteration is planned.

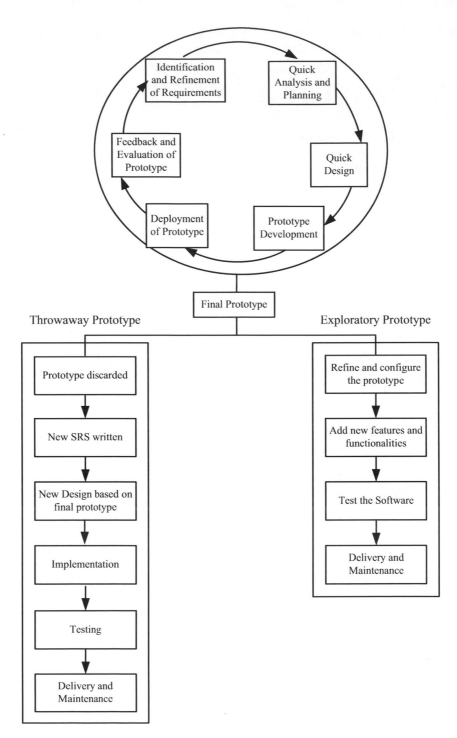

FIGURE 1.6
The prototyping development paradigm.

Depending on the implementation method and the customer requirements, there are two approaches to prototyping: throwaway prototyping and exploratory prototyping.

i. **Throwaway prototyping**

Once the development team and the customer finalize the prototype, it is put to one side. Taking the final prototype as the basis for the model, a new SRS with detailed configurations is written. A rapid new design is built using new tools and technologies. The finalized prototype is discarded and exactly the same operational software is developed. The discarded prototype works as a dummy guide used to refine the requirements, as shown in Figure 1.6.

ii. **Exploratory prototyping**

In exploratory prototyping, the finalized prototype is not discarded; rather it is treated as the semi-finished product. New features and additional functionalities are added to make the prototype operational. After reviews by the customer and testing by the developer it is deployed to the customer's site.

1.4.3.1.1 Key Findings

- Prototyping is the most suitable paradigm for eliciting requirements from the customer. Non-technical customers may be unaware of or uncertain about their software requirements; after seeing working or dummy prototypes of similar products, customers can explore more specific requirements.

- Once the development team and customer finally agree on a particular prototype, the development of the final product takes minimum time compared with other software development paradigms.

- Analysis, design or coding errors by the development team can be identified early in the development phases, and can be rectified without delay. It improves the overall life cycle of the software.

- Of all the traditional development paradigms, customer involvement is at its highest level in the prototyping model. In all iterations, customer feedback is considered and incorporated in the next prototype iteration.

- Prototyping is one of the most flexible development paradigms in terms of incorporating changes or extensions to the requirements.

1.4.3.1.2 Critiques

- One of the major criticisms of the prototype paradigm concerns long-run quality issues. In the rush of delivering operational software, quality is sometimes compromised. To fulfill the customer's demands as soon as possible, inefficient, incompetent or poor programming languages can be chosen or inappropriate databases selected, or the developer may choose technologies for their own convenience rather than to meet the demands of the proposed software.

- If the customer's demands change too frequently, then the time taken to develop the prototype is longer than the time to develop the actual software.

- Sometimes, it is difficult to convince other stakeholders about the prototype development paradigm. If after a while the customer is informed that the product developed so far is only a prototype, they may raise objections regarding the time and cost invested.

- For large and complex projects, the prototype model may become unmanageable.

1.4.3.2 The Spiral Development Paradigm

The spiral paradigm is probably the most famous traditional development paradigm. It was one of the first paradigms to include the assessment of "risk factors" involved in the proposed software. The spiral paradigm combines the features of prototype and classic sequential development paradigms in an iterative fashion. The spiral model consists of loops. Each 360° loop denotes one complete cycle. The spiral paradigm consists of four quadrants:

Quadrant 1: Define objectives, identify alternative solutions and identify constraints.

Quadrant 2: Evaluate alterative solutions, identify possible risks, mitigate identified risks.

Quadrant 3: Development, construction and verification of software.

Quadrant 4: Planning of the next cycle.

After the completion of each spiral cycle, some milestones are achieved and new milestones are established (Figure 1.7).

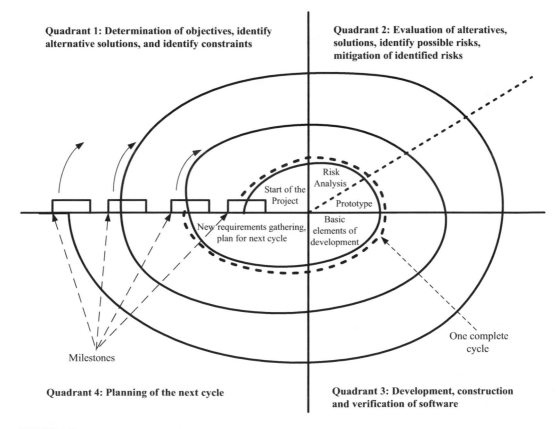

FIGURE 1.7
The spiral development paradigm. (Adapted from Boehm 1986.)

1.4.3.2.1 Key Findings

- The spiral paradigm is especially useful when the cost factor as well as the risk factor in the software is high. It assesses the risk involved in the software in each iterative cycle. Risk assessment early on in development may save many final-stage unforeseen events, as any risk can delay the schedule or even take the software over budget.

- The spiral paradigm is one of the most interactive development paradigms, with the involvement of the customer highly appreciated and encouraged. Customers are involved in almost every phase of the life cycle of their product. They cannot provide inputs but they are entitled to review the product and can record their feedback.

- The spiral paradigm provides flexibility in terms of adding and/or changing requirements in every cycle. New requirements or changes in the previous requirements can be easily accommodated in the next development cycle.

- The spiral paradigm is a never-ending paradigm. Even after the delivery of the software, it is still adaptable and can accept changes, modifications and enhancements in the running software.

1.4.3.2.2 Critiques

- To assess and mitigate the risk, experts are required. If risk is not evaluated properly or identified in time, it may generate serious consequences.

- It is not suitable for small-scale projects. The spiral paradigm requires a sufficiently large amount of both time and cost to plan, review and assess each cycle, which may not be cost-effective for small projects.

- The spiral paradigm may generate a huge amount of paperwork, as each phase consists of various changes, prototypes and similar intermediate versions.

- Sometimes it is hard not only to establish realistic milestones but also to achieve them, as the spiral paradigm supports continuous extensions and changes in the software.

- It is a comparatively costly development paradigm, due to the involvement of risk assessment features, the building and evaluation of prototypes in each cycle, and continuous changes suggested by the customer.

1.5 Advanced Software Engineering Paradigms

The field of software engineering evolves with time and experience. Millions of software engineers, industry experts, researchers and academics are involved in the process of providing better and efficient solutions to real-world dynamic problems. It is their efforts that have led to the evolution of advanced software engineering paradigms. In this section, we discuss some well-known software engineering paradigms that have appeared in recent years and follow the different approaches to development. Many advanced development paradigms are available in the literature, but here we focus on four specific paradigms: agile, component-based development, aspect-oriented development and cleanroom development (Figure 1.8).

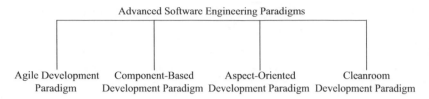

FIGURE 1.8
Advanced software engineering paradigms.

1.5.1 Agile Development Paradigm

The agile development paradigm is actually a combined development approach which includes a set of guidelines, self-organizing and motivated teams, and collaborations with customers. The sole aim of the agile paradigm is customer satisfaction. It emphasizes continuous planning that can adapt to changes, evolutionary and iterative design and development, continual improvements, and rapid and flexible response to change. In 2001, Kent Beck and a team of 16 eminent developers and researchers known as the Agile Alliance published their "Manifesto for Agile Development" (Beck et al. 2001), the fundamental guide to the new and innovative agile development paradigm. The "Manifesto" defined four common development values:

i. "Individuals and interactions over processes and tools," that is, customers and interactions with customers will be given preference over development processes and tools.

ii. "Working software over comprehensive documentation," that is, a focus on giving working and operational deliverables rather than generating documents.

iii. "Customer collaborations over contract negotiations," that is, the customer is involved in the project and developers have ownership of the software rather than just developing and delivering the software to the customer.

iv. "Responding to change over following a plan," that is, changes in the software at any stage are welcomed and adaptations made rather than waiting for the next phase after the review, as is the case in traditional development paradigms.

To attain and realize the agile paradigm in the best way, the "Manifesto for Agile Development" describes the 12 principles of agility (Agile Alliance 2003, Fowler and Highsmith 2001), as:

i. "Highest priority of agile paradigm is to satisfy the customer through early and continuous delivery of valuable software."

ii. "Agile paradigm welcomes changing requirements, even late in development. Agile processes harness change for the customer's competitive advantage."

iii. "Deliver working software frequently, from a couple of weeks to a couple of months, with a preference to the shorter timescale."

iv. "Business people and developers must work together daily throughout the project."

v. "Build projects around motivated individuals. Give them the environment and support they need, and trust them to get the job done."

vi. "The most efficient and effective method of conveying information to and within a development team is face-to-face conversation."

vii. "Working software is the primary measure of progress."

viii. "Agile processes promote sustainable development. The sponsors, developers, and users should be able to maintain a constant pace indefinitely."

ix. "Continuous attention to technical excellence and good design enhances agility."

x. "Simplicity, the art of maximizing the amount of work not done, is essential."

xi. "The best architectures, requirements, and designs emerge from self–organizing teams."

xii. "At regular intervals, the team reflects on how to become more effective, then tunes and adjusts its behaviour accordingly."

Agility is a philosophy rather than just a development paradigm. It is not about simply following the defined plan or predefined activities and operations. The agile paradigm defines the process of development as a collaborative endeavor among stakeholders of the software, self-organizing and motivated teams working in a healthy environment to achieve a common goal: customer satisfaction.

In the agile development paradigm, large and complex software is divided into smaller and manageable increments to avoid the rush of delivering the full-fledged software at the final stage. Quick design and rapid implementation are performed by expert teams making the deliverable available for customer review.

The agile paradigm is actually a methodology, a philosophy of development. There are various frameworks available to implement this methodology. Some of the best-known frameworks are;

- Scrum
- Extreme Programming (XP)
- Crystal
- Lean Software Development (LSD)
- Adaptive Software Development (ASD)
- Dynamic Systems Development Method (DSDM)
- Feature Drive Development (FDD)

1.5.1.1 Key Findings

- Unlike traditional paradigm teams consisting of analysts, designers and developers, in the agile development paradigm, the team consists of all the stakeholders of the software, that is, the customer or the customer's representative is now an integral part of the development team.
- "Customer's stories" are replaced by "customer's voice." Rather than documenting the requirements of the customer, here the customer is involved in the elicitation, planning and analysis of the requirements as a member of the development team.
- "Customer satisfaction" is at the heart of the agile development paradigm. If the customer is satisfied with the deliverable product, its cost and the schedule, then the agile paradigm is on the right track.

- The agile paradigm concentrates on early delivery of operational software in increments, discarding intermediate prototypes and documentation that are not useful to the customer.

- Changes are always welcomed and adaptations enabled by the agile team at every stage, even during the late phases of development.

- Customers feel satisfied as they not only get operational software quickly, but they feel like members of the development team.

- Implementation and testing run in parallel, which ultimately greatly reduces time schedule and overall cost of the software. In addition, errors can be caught early on in development and rectified without delay or before they generate serious consequences.

- At any stage of development, if a decision is taken to rollback, only that particular increment will be affected.

- The agile development paradigm is helpful for both types of customer: those who clearly know their requirements and the scope of the final product; and less well-informed clients who have no idea of their software requirements.

1.5.1.2 Critiques

- All the emphasis is on customer satisfaction, adaptability to change and developer skills, but there is little focus on developers' or software engineers' motivation or, in Alistair Cockburn's words, "the agile alliance forgets the frailties of the people who build computer software" (Cockburn 2002).

- While delivering rapid increments, sometimes it is difficult to assess the actual effort and resources required for large and complex software.

- With the main emphasis on adaptation to changes suggested by the customer, implementation of those changes, and rapid delivery, detailed design and detailed documentation may be neglected.

- If the development team or the customer is not clear about the scope of the requirements, or the customer is not satisfied with the increments and makes changes quite frequently, agile may become an unmanageable or even interminable process.

- Agile demands experts and experienced designers and developers in the development team, who can design and implement new or changed requirements efficiently and as quickly as possible. The success of the agile development paradigm depends to a considerable extent on the availability of these domain experts.

- For low-budget or small-scale projects, agile is not the right development paradigm.

1.5.2 Aspect-Oriented Development Paradigm

As the size of the software increases, so too do cost, schedule and complexity. There are basically two reasons why size and complexity of the proposed software increase:

1. One of the major reasons for size increments is the repetition of functionalities in every component or module of the software. Sometimes it is a requirement of the application that some functions are required in all components of the software, such as "memory management," "logging," "resource sharing," "error handling," "exception handling," "business rule," "transaction processing," and "real-time constraints." The problem

arises when these repeated functions require regular changes and updates to every module in which they appear. Because of the increment in complexity and size these small function codes are neglected, causing problems in the software.

2. Sometimes adding or modifying a single functionality/module/component requires almost all the software modules to be changed.

In the aspect-oriented paradigm, these elements are called "aspects" or "concerns." The aim of aspect-oriented software is to modularize or isolate these complex "crosscutting" concerns or aspects. The aspect-oriented development paradigm is a set of processes and techniques for identifying, defining, specifying, designing and constructing "aspects." "Aspect oriented development paradigm is the mechanisms beyond subroutines and inheritance for localizing the expression of a crosscutting concern" (Elrad, Filman and Bader 2001).

Common terms used in the implementation of the aspect-oriented paradigm are as follows:

- **Aspect:** From the implementation point of view an aspect can be a module or a class that contains crosscutting concerns from different modules.
- **Joinpoint:** Joinpoints are the well-implemented points in the program that act as a junction of two functionalities or functions, such as a method, the entry point of a method, the exit point of a method, throwing an exception, and so on.
- **Advice:** Advice is an action or set of instructions that are executed on a joinpoint and initiated by an aspect.
- **Pointcut:** Pointcuts are groups of joinpoints that are checked before an advice is executed.

1.5.2.1 Key Findings

- Aspect-oriented development helps to isolate core concerns from other concerns. It is helpful to identify important and fundamental areas of code where crosscutting aspects can be identified and designed.
- Aspect identification and design is not limited to the aspect-oriented paradigm; rather, it can be used with any other paradigm to enhance the efficiency of the code.
- It helps to modularize the overall architecture of the design which ultimately helps to maintain the software.
- This paradigm appreciates the reuse of coding constructs, so that the time and cost of development can be sufficiently reduced.
- It complements and extends the object-oriented development paradigm concepts.
- It promotes the philosophy of "separation of concerns."

1.5.2.2 Critiques

- It is difficult to identify and define concerns and aspects that actually assist with crosscutting or modularization, hence an expert team is required.
- Since it is assumed that this paradigm is still evolving, many concepts and concerns remain unanswered in the paradigm.

- If concerns are not specified and designed well, this paradigm may create overheads. Even where aspects are well designed, it is still a time-consuming process to write, track, modify and maintain concerns.

- For small-scale applications it is easy to specify and design concerns, but for large and complex software it is very difficult to manage aspects.

- Not much documentation is supported by the aspect-oriented paradigm; even these concerns and aspects are in practice limited to coding.

1.5.3 Cleanroom Development Paradigm

Cleanroom is a formal development paradigm. It is a different method of software engineering based on formal methods rather than the code-and-test methods normally followed by traditional development paradigms. The cleanroom paradigm emphasizes mathematical proofs and mathematical foundations. It provides a certain level of software reliability from the outset. The fundamental attribute of this paradigm is "defect prevention" rather than "defect detection and then removal."

The philosophy behind cleanroom software engineering is to avoid dependence on costly defect removal processes by writing code increments correctly the first time and verifying their correctness before testing. Its process model incorporates the statistical quality certification of code increments as they accumulate into a system. (Linger 1994)

The term "cleanroom" has previously been used in electronics to mean precluding the inclusion of errors and faults in semiconductors during manufacture. In the context of software development, "cleanroom" expresses the property of correctness in the developed software. Cleanroom provides correctness through formal specifications, walk-throughs, formal verification and inspection methods.

Cleanroom is incremental in nature, that is, it produces software in small increments rather than developing and delivering the complete software all at once. Each increment is properly verified, tested and certified by the development teams. Further, these increments are assembled to construct the complete software. The phases involved in each increment of the cleanroom paradigm are shown in Figure 1.9. Each increment consists of:

- **Planning:** Planning is undertaken with special care, such that available increments can be easily adapted and integrated into the current increment.

- **Requirement analysis:** Requirements can be collected from any traditional or advanced methods and tools.

- **Box-structure specification:** Specifications are defined using box structures. According to Hevner and Mills (1993), box structures "isolate and separate the creative definition of behavior, data, and procedures at each level of refinement." Three types of box structure are used to specify the software in the cleanroom paradigm:

  1. *Black box* specifies the behavior of the system when a function is applied to the set of inputs. Here control is focused on inputs provided and outputs received.

  2. *State boxes* are similar to encapsulated objects that put input data and transition operations together. They are used to state the states of the system.

  3. *Clear box* describes the operations that are transmitting the inputs to the outputs.

- **Formal design and correctness verification:** Designs are achieved using formal specification methods in addition to traditional design methods. Formal design methods

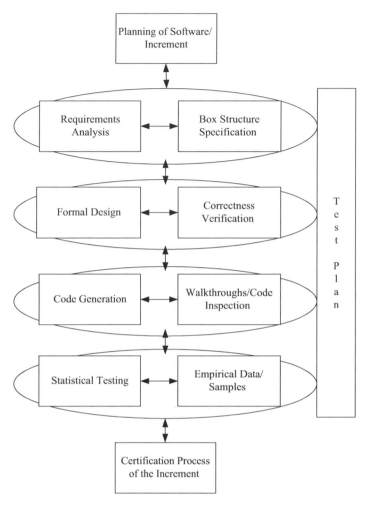

FIGURE 1.9
One increment of cleanroom software development.

are used at architectural as well as component level. The cleanroom paradigm regressively focuses on verification of the design. Mathematical proofs and logics are used for the correctness of the proposed design.

- **Code generation, inspection and walkthroughs:** The cleanroom paradigm translates the corrected design into appropriate programming languages. Code inspections, decisional charts and technical reviews are conducted to check the correctness of the code.

- **Statistical testing with empirical data:** In the cleanroom software development paradigm, testing is conducted using statistical data sets and samples. Testing methods are not only standard but also mathematically verified. Metrics support mathematical validations. Exhaustive and regression testing is performed regressively.

- **Certification of increments:** Once all the phases are tested and verified, the certification of the increment by the development team begins. All the necessary tests are conducted and the increment is released so that it can be integrated with other certified increments.

1.5.3.1 Key Findings

- In the cleanroom development paradigm, test plans are prepared just after the requirements have been collected and continuously throughout the development phases. It focuses on error prevention from the beginning.

- Regressive correctness verifications and code inspections are performed on the design and coding of the proposed software to provide a certified level of reliability to the software.

- The mathematical basis of the paradigm makes the developed software less prone to small errors.

- It reduces the effort as well as the cost of testing, which occurs in almost the final stages of traditional development paradigms.

- The use of box structures makes the paradigm more reliable.

1.5.3.2 Critiques

- To apply the cleanroom software paradigm, trained and experienced people are required who have a clear understanding of formal methods of development.

- It is a time-consuming methodology compared with other paradigms, since formal specification, inspections, peer review and statistical testing take a considerable time.

- It can be difficult to convince customers of how this paradigm works. The concepts of mathematical proofs and box-structure representations are hard to understand.

- It is comparatively less practiced by software engineers due to its complex structure and mathematical verification methods.

1.5.4 Component-Based Development Paradigm

The component-based development paradigm is now an established field of software engineering which from the outset focuses as much as possible on reusability of pre-constructed code constructs and other usable or reusable deliverables. The component-based paradigm can be utilized with both types of development paradigm, traditional or advanced. In component-based paradigms, reusability can be at any level, or with any deliverable, whether in requirements, documentation, design construct, coding, test cases or any similar software construct.

1.5.4.1 Key Findings

- In the component-based paradigm, the emphasis on reusability ultimately results in a shorter development life cycle. Component-based development supports modularity, or componentization, which helps control the overall design of the software and ultimately provides better maintainability to the developed software.

- Due to the nature of componentization, the component-based paradigm supports parallel construction of components to increase productivity in the development.

- The component-based paradigm supports reusability not only during but also after development. Components are deposited in the repository for future use.

- Reduced development costs in comparison to other development paradigms.
- Components can be added or removed without affecting the other parts of the software, hence it provides better scalability.
- It enhances the overall reliability of the software as pre-tested and properly implemented components are used in development.
- It is assumed that components used in development support language independence and vendor independence.
- It provides greater flexibility in terms of development and user interaction.

1.5.4.2 Critiques

- Component-based development supports assembly of pre-developed components. Assembly and integration is a major challenge in this paradigm. No standard mechanism has been defined for the integration of components into the software architecture. The only mechanism available is the design document.
- Interaction among components is both crucial and time-consuming. Interaction issues are complex and generate complexity in the overall design of the software.
- Testing is comparatively time-consuming and costly as the concept is based on integration and interaction of components. The main focus is on integration testing of components.
- Components can be sourced from different vendors or from a third party. Component quality may therefore vary, ultimately affecting the overall quality of the developed software.
- Experts are required to identify and integrate components according to the architectural design and requirements of the software.

For further discussion of the component-based development paradigm, see Section 1.6.

1.6 Component-Based Software Engineering (CBSE)

Component-based software engineering (CBSE) is an elite form of software engineering that offers the feature of reusability. Reuse of software artifacts and the process of reusability make CBSE a specialized software development paradigm. Its philosophy is "buy, don't build." CBSE reuses pre-constructed and available software constructs rather than developing them from the beginning (see Figure 1.10). The basic idea is to develop a component (including classes, functions, methods and operations) only once and reuse it in various applications rather than re-constructing it every time. Reusing pre-developed and pre-tested components make the development life cycle shorter, helps to increase the reliability of the overall application and reduces time to market.

According to Gaedke and Rehse (2000), "Component-Based Software Development (CBSD) aims at assembling large software systems from previously developed components (which in turn can be constructed from other components)." Component-based software applications are constructed by assembling reusable, pre-existing and new

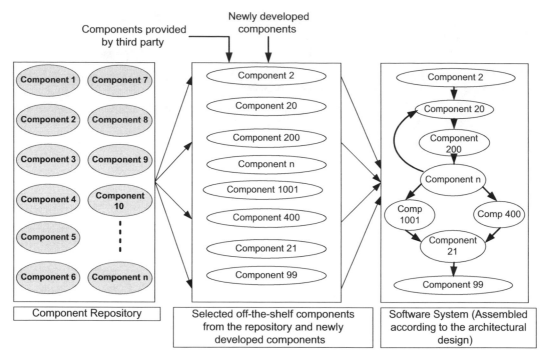

FIGURE 1.10
Component-based software engineering framework.

components which are integrated through error-free interfaces (Shepperd 1988, Heineman and Councill 2001, Lin and Xu 2003, Capretz 2005, Jiangou et al. 2009). The objectives of CBSE are to develop extremely large and complex software systems by integrating commercial off-the-shelf components (COTS), third-party contractual components and newly developed components to minimize development time, effort and cost (Boehm et al. 1984, Brereton and Budgen 2000). CBSE offers an improved and enhanced reuse of software components with additional properties including flexibility, extendibility and better service quality to meet the needs of the end user (Basili and Boehm 2001, Jianguoet al. 2011, Vitharana 2003, Kirti and Sharma 2012).

The CBSE development paradigm is used to develop generalized as well as specific components. Four parallel processes—new component development, selection of preexisting components from the repository, integration of components and control and manage—are involved in the creation of more than one application concurrently, as described in Figure 1.11. With each process there must be a feedback method to address problems and errors arising in component selection, new component development, interaction and integration errors among the components, and their side effects. To manage all these parallel activities, there must be a control procedure or management procedure which will not only assess the development process but will also manage the requirement analysis, selection of components, integration of components and, most importantly, the quality of components submitted to the repository for future reuse.

Jacobson, Griss and Jonsson (1997) define four similar concurrent processes associated with development of various applications in CBSE. These processes are creation, management, support and reuse. The creation process involves developing new applications

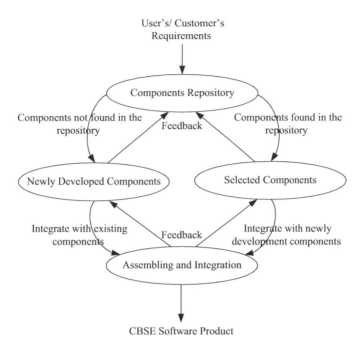

FIGURE 1.11
Concurrent component-based software development processes.

by reusing existing software components. The management process manages the activities of the development, including selection of components according to requirements, cost of component selection and schedule of new application development. The support process provides help and maintenance activities in the development of new applications and provides existing components from the repository. The reuse process collects the requirements and analyzes them to select components from the repository, and is responsible for the actual development of components through the property of reusability.

1.6.1 Evolution of Component-Based Software Engineering

The idea of components was introduced by Douglas McIlroy in a defining speech titled "Mass-Produced Software Components" given at the NATO conference on software engineering in 1968 (McIlroy 1976). In the 1980s and 1990s the term "component" acquired the identity of a building block and architectural unit. Some 40 years on, CBSE has become established as a standard software engineering paradigm and one of the most promising and capable paradigms for the development of large, complex and high-quality software systems.

The evolution of CBSE can be divided into four phases: the preparatory phase, the progression phase, the defined phase and the expansion phase.

1.6.1.1 Phase 1: The Preparatory Phase

In its early stages, component-based software development researchers collected information about CBSE. Worldwide conferences and workshops were organized to initiate and define approaches, challenges and implications of CBSE. The first workshop on CBSE was held in Tokyo in 1998. The major achievements of the initial phase are as follows (Heineman and Councill 2001):

a. CBSE became a popular software development paradigm.

b. Component was assumed and defined as a pre-existing and reusable entity.

c. Standard CBSE literature was produced, including Heinemann and Councill's *Component-Based Software Engineering: Putting the Pieces Together* (2001).

1.6.1.2 Phase 2: The Defined Phase

In the second phase, practitioners focused on more specific and specialized branches of CBSE. Terms relating to CBSE were specifically defined and classified. Worldwide conferences and workshops were organized and new research fields and research topics like "predictable assembly," "component trust," "component specifications," and "automated CBSE" came into existence. The milestones of the defined phase are as follows:

a. Specific and concrete terms in the context of CBSE were defined, including "component architecture," "component specification," "component adaptation," "component acquisition," and so on.

b. These defined topics created new research areas and research groups.

c. The number of academic and research publications increased and organizations such as ACM started to publish in the area of CBSE.

1.6.1.3 Phase 3: The Progression Phase

In the third phase, new and inventive areas of CBSE began to emerge. In this period not only research groups but also the software industry began to address the benefits of CBSE. The major CBSE research areas during this phase are component-based system modeling and design, component testing framework, component concepts for embedded structures, estimation of resources at dynamic time in multi-task CBS environments, design and analysis of component-based embedded software, component structure for critical dynamically embedded contexts and the hierarchical framework for component-based real-time systems. This phase is characterized by the collaboration of the component-based paradigm with other areas in software engineering. The recognized progress of this phase is that new and collaborative areas have been identified and defined, such as multi-task, real-time systems, embedded systems, critical systems and performance attributes, in collaboration with CBSE.

1.6.1.4 Phase 4: The Expansion Phase

It is assumed that this phase is still continuing. This is the time in which parallel and balancing approaches are recognized in software engineering and software development. Areas like service-oriented development, aspect-oriented programming and model-driven engineering are running in parallel with CBSE. The major research areas in this phase are component-based web services, service-oriented architectures, development, customization and deployment of CBSE systems, components supporting grid computing, global measurement, prediction and monitoring of distributed and service components, integrated tools, techniques and methods for constructing component-based service systems, components for networked dynamic and global information systems-linking sensor.

1.6.2 Characteristics of Component-Based Software Engineering

Heineman and Councill describe three fundamental properties of CBSE: (i) the development of systems from existing software entities; (ii) the capability to reuse these pre-existing entities as well as newly developed software entities in other applications; and (iii) convenient preservation, maintenance and fabrication of these entities for future implementations. CBSE possesses a number of properties over and above those of other software engineering paradigms. Some primary characteristics of CBSE are listed below.

- **Reusability:** Reusability is the focal property of CBSE. Krueger (1992) defines software reusability as "the process of creating software systems from existing software rather than building them from scratch." Software reuse is the process of integrating predefined specifications, design architectures, tested code, or test plans with the proposed software (Johnson and Harris 1991, Maiden and Sutcliffe 1991, Berssoff and Davis 1991). CBSE relies on reusing these artifacts rather than re-developing them. Components are developed in such a way that they can be heterogeneously used and then reused in various environments.

- **Composability:** One of the fundamental characteristics of CBSE is that the components are the reusable, composable software entities, that is, applications are composed of different individual components. These individual reusable components are designed in such a way that they can be reused in composition with other components in various applications with minimum or no fabrication. Components are composed of components, that is, a component is itself made up of components, which are further made up of other components, and so on (Atkinson et al. 2002). A component can be a part of one or more components.

- **Shorter development cycle:** The component-based development paradigm follows the "divide, solve and conquer" approach. In this paradigm, complex and bulky applications are divided into smaller, more manageable units or modules. Then, rather than starting coding of a complete module from the first line, existing elements are sought and assembled that satisfy the requirements of the module under consideration. This increases software development speed. In addition, several modules can be implemented concurrently, regardless of location or context. Thus, development time is saved and the development cycle becomes shorter.

- **Maintainability:** The effort required to add new functionalities to the application, or modify, update or remove old features from the software, is referred to as the maintenance. Since CBSE-based software is made up of reusable and replaceable components, we can add, update, remove or replace components according to the requirements of the software. Maintaining composable and independent components is much easier than maintaining monolithic software.

- **Improved quality and reliability:** Component-based technology ensures superior quality as the CBSE integrates and couples pre-tested and qualified components. These components are specifically tested at least at the unit level. During their integration with other pre-tested components, the developer performs integration as well as system tests (Brown and Wallnau 1998). This regressive form of testing makes component-based applications more robust and improves the quality of the product. More broadly, the effort, cost and time of testing are noticeably reduced. Components are independently developed, deployed in various contexts at the same time with minimal or no fabrication and integrated according to the predefined architecture; hence, it is assumed that there are no unwanted interactions among the components to make them unreliable. All the interaction paths are predefined so the reliability and predictability of components is increased.

- **Flexibility and extendibility:** Software developers have the choice of customizing, assembling and integrating the components from a set of available components according to their requirements. Replaceable and composable components are easy to add, update, modify or remove from the application without modifying other components. Error navigation and fixing are relatively easy as it is limited to component level only.

1.6.3 Componentization

Componentization is the method of identifying the quantity of components in a specific application developed through component-based development. Componentization addresses the issue of maintaining a balance between the number of components and the complexity factors of the system. Essentially, the level of componentization is equivalent to the level of requirement sufficiency. In determining the level of requirement sufficiency, we consider as many components as are adequate to solve the software application's intention. Figure 1.12 illustrates componentization vs integration cost.

If we divide the problem into a large quantity of components providing small functionalities, it will increase both the cost of integration and the interaction effort. Not only the cost but also the number of interactions, the coding complexity, testing effort and number of duplicate test cases will also increase. If an application is componentized with fewer components each providing a number of functionalities, it will cost in terms of testing as well as maintenance. It is desirable to achieve a minimum cost region so that cost and effort can be balanced against the number of components.

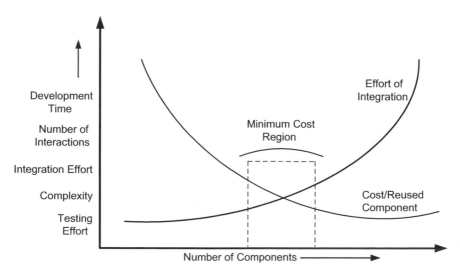

FIGURE 1.12

Componentization vs integration cost. (Adapted from Pressman 2005.)

1.7 Components

"Component" is defined by various dictionaries as follows:

> Contributing to the composition of a whole (*Oxford English Dictionary*, 2018).

> A component is any part of which something is made (*Oxford Dictionary*, 2018).

> A constituent part; an ingredient (*Webster's Dictionary*, 2018).

> A part which combines with other parts to form something bigger (*The Cambridge Dictionary*, 2018).

The core element of CBSE is the component itself. Components are the basic unit of reusability and the building blocks of component-based software. They are the focal point and prime operative part of the component-based development paradigm. In the literature, components have been defined by various researchers in a variety of ways. Some leading definitions are:

Definition 1. Szyperski and Pfister: "A software component is a unit of composition with contractually specified interfaces and context dependencies only. A software component can be deployed separately and is subject to composition by third parties" (Szyperski and Pfister 1997).

Definition 2. Booch: "A component is a logically cohesive, loosely coupled module" (Booch 1987).

Definition 3. OMG: "A component is a modular, deployable, and replaceable part of a system that encapsulates implementation and exposes a set of interfaces" (Object Management Group 2000).

Definition 4. Brown: "A software component is an independently deliverable piece of functionality providing access to its services through interfaces" (Brown 2000).

Definition 5. Heinmann and Councill: "A software component is a software element that conforms to a component model and can be independently deployed and composed without modification according to a composition standard" (Heineman and Councill 2001).

Definition 6. Szyperski: "Software components are binary units of independent production, acquisition and deployment that interact to form a functioning system" (Szyperski 1999).

Definition 7. D'Souza and Wills: "A coherent package of software artifacts that can be independently developed and delivered as a unit and that can be composed, unchanged, with other components to build something larger" (D'Souza and Wills 1998).

Definition 8. Sparling: "A component is a language neutral, independently implemented package of software services, delivered in an encapsulated and replaceable container, accessed via one or more published interfaces" (Sparling 2000).

Definition 9. Kruchten: "A software component is a nontrivial, nearly independent, and replaceable part of a system that fulfils a clear function in the context of a well-defined architecture. A component conforms to and provides the physical realization of a set of interfaces" (Kruchten 2003).

Definition 10. Hopkins: "A software component is a physical packaging of executable software with a well-defined and published interface" (Hopkins 2000).

Definition 11. Meyer: "A component is a software element (modular unit) satisfying the conditions: a) it can be used by other software elements, b) it possesses an official usage description, which is sufficient for a client author to use, and c) it is not tied to any fixed set of clients" (Meyer 2003).

From an analysis of these prominent and diverse definitions, one can deduce that they all contain a number of universal points. In the context of this work and in the light of these universal directives, we can identify a general definition of components as follows:

- A component is an identifiable, functionally reusable unit (element/part/piece/package) of a larger software system.
- A component is developed independently.
- A component is delivered and deployed in a context-free environment.
- A component interacts with other components through well-defined interfaces to access and to provide services.
- A software component can be a replaceable, modifiable piece of code that can be integrated with other replaceable, modifiable components.

1.7.1 Types of Components

A component is an identifiable and functionally reusable unit that can be reused at various levels with different degrees of reusability. Bennatan (2003) describes three classes of component:

- **Off-the-shelf components:** In this category, components are pre-developed, pre-tested and can be reused without any modification. Off-the-shelf components are either made available by the third party or selected from the repository. Components residing in the repository are constructed by the developers of the organization as part of previously developed software. Off-the-shelf components are like black-box components, as the developers can use or reuse them without reference to the idiosyncrasies and internal details of that component. As their code and internal structure are not available, interfaces of such components should be self-descriptive, that is, their interaction details must be defined in their interfaces. These components can be deployed in a context-free environment without making any change or modification. Off-the-shelf components are also known as commercial off-the-shelf (COTS) components.

  - **Adaptable components:** Components in this category are assembled with existing software artifacts including application requirements, architectural design, available code or pre-built test cases with small or large modifications. We can divide adaptable components into two broad classes: *fully qualified components* and *partially qualified components*. This division is based on the degree of modifiability of the component. Fully qualified components need no alteration or a minimum amount of alteration. They can be reused at their maximum extent with minimum modification. The modification requirement depends on the current application requirement. Partially qualified components need a greater degree of alteration. These components can be reused in the application under consideration with major modifications (Tiwari and Kumar 2016).

  - **New components:** New components are those that are engineered by the organization's developers from the first line of code to the last, considering the requirement specification of the specific software application. New components are developed when neither off-the-shelf nor adaptable components satisfy the requirements of the software system. In CBSE, components are developed in such a way that they can be reused in various applications. Newly developed components are submitted to the repository for future use.

Components can be developed independently, delivered and deployed in a context-free environment. Components interact with each other to gain and to offer their services. In component-based development, components interact and communicate to contribute their services to the software system (Gill and Balkishan 2008). When components communicate with each other, they share information including data, logic and other valuable assets (Basili and Boehm 2001, Jianguo et al. 2011).

1.7.2 Characteristics of Components

To gain as well as to provide valuable information, components require three fundamental properties: *interfaces*, *services* and *deployment techniques*.

- **Interface:** To provide and to access information and services, components require some sort of intermediary between them, that is, an interface. An interface is a medium which defines how a component can interact with other components. It states the communication structure, limits and parameters and also describes the form and structures of the result. In CBSE, developers use interfaces to assemble and create interactions among the dependent and independent components. Interfaces define the mode of coupling between two or more components. The signature (data type, count) of corresponding parameters and the nature of the communication among the components decide the type and mode of coupling. Components use interfaces to pass the control sequence from one component to others, to acquire and provide files and data, and to transmit instructive information to each other. When components are integrated through various coupling techniques to contribute their services and functionalities to the whole system, the interface is the medium that provides the method of interaction among these components (Pressman 2005).

- **Services:** To survive in a system, every component must have to provide some desired and defined service (set of services) or functionality (set of functionalities). In component-based software systems, the service specifies a component's role in the system. When developers select a component from the repository, one of the basic selection criteria is the service that it will provide to other components and to the system. In addition, when components are stored in the repository, the basic aim is to reuse their services in future applications. A component's service consists not only of its functional behavior but also of the functional procedures that would be assembled with other pre-existing components (Schach 1990). In the context of CBSE, it is clear that
  - Services must have some specific and defined purpose,
  - Services should exhibit robust behavior,
  - Services should be reliable,
  - Services should exhibit efficiency in terms of performance and adaptability.

- **Deployment Techniques:** In component-based development, components are deployable units. Therefore, it is necessary to have a deployment technique for every component. Deployment techniques are the methods through which a component is customized for the purpose of its implementation. If component's source code is available, customization is achieved through its executable code, or otherwise through its interface (Nadeesha 2001). Deployment techniques conform to the syntax and semantics of the language in which the component was developed. In the context of black-box and context-free components, generally internal details are not available with the component. In such cases, interfaces play a major role in the deployment. To make deployment effective, interfaces must be robust and reliable.

When a large and complex system is to be developed, it is divided into comparatively small manageable pieces. In component-based systems, components represent these small pieces. Components are replaceable, modifiable pieces of code that can be integrated with other replaceable, modifiable components. To address the requirements of the system, it is essential to partition and modularize these requirements. Modularization must be done in such a way that each component represents a specific requirement (or set of requirements).

1.7.3 Component Repository

In CBSE, reusability not only applies at the time of development but is also maintained after development. When components are developed by the development team, developed/contracted by the third party or purchased, they must be stored in the repository for future use. The repository is the component database which contains the following:

a. Coding of the component

b. Metadata of the component

c. Components interface

d. Other information about the component.

Repositories need to be regularly maintained to ensure their consistent efficiency. Repository maintenance is one of the major concerns in CBSE. Some major issues associated with component repositories are as follows:

a. Identifying useful and outdated components to update the repository.

b. Updating components, that is, elimination of outdated components and inclusion of new components in the repository.

c. Updating component metadata and other information.

d. Keeping different versions as well as updating versions of the same components without duplication or ambiguity.

1.7.4 Component Selection

The selection of appropriate components for the proposed software is crucial in many respects. Proper and accurate selection of components makes the software economic, helps to shorten development time, increases reliability and makes the overall development much smoother. However, in order to select one component from a number that are available with similar intentions, selection and verification criteria are needed. At present the only criteria available for component selection are the customer requirements. Developers generally look for a component that fits into the software design without considering other selection parameters. Apart from the requirements, major issues in component selection are:

a. Selection based on the level or degree of reusability of components.
b. Selection based on the reliability issues of the component.
c. Selection based on the increased/decreased level of complexity in the overall software.
d. Selection based on the component testing efforts.
e. Selection based on the scope of maintenance of the component.
f. Selection based on the scope of scalability in the component.

1.7.5 Component Interaction

In component-based software development, interaction among components is necessary as they are integrated to provide features and functionalities for the proposed software. Components are designed to interact with each other according to the requirements of the software. Component interaction is not only necessary but plays a crucial role in the design of the software architecture. Components are assumed to be context independent, language independent and vendor independent. Researchers have used various methods, including graph theory notation, to show the interaction among components. UML includes component interaction diagrams to denote such interactions.

Some fundamental issues with the interaction of components are as follows:

a. Interactions among components generate complexity in the software. The aim is to reduce interaction complexity among components.
b. It is necessary to define the interaction mechanism, parameters or methods among components.
c. There is a need for common interfaces to make an interaction between two components.
d. Interaction of components invokes integration-related issues including integration testing and regression testing.
e. Interaction can introduce new software errors which may increase the testing and debugging schedule and cost.
f. Appropriate and well-defined metrics are needed to assess these interactions so that the contribution of components can be justified.

1.7.6 Component Dependency

Interaction and integration among components create dependencies among them. Components depend on each other for functionalities and services. Component dependency governs the level of coupling among the components. As the dependency increases, interaction among components will also increase. These interaction dependencies generate complexity in the software architecture. Components can be dependent on each other for various reasons, including:

i. Dependency due to data sharing
ii. Temporal dependency

 iii. Dependency due to controlling the flow of execution

 iv. Sharing of the same state or state dependency

 v. Dependency due to providing input or due to receiving output

 vi. Context dependency

 vii. Interface dependency

Major issues in the context of component dependencies are as follows:

a. Dependency among components should be kept to a minimum if it cannot be elimi-nated completely.

b. Notation of dependency in the design of software should be unambiguous.

c. Presentation of dependency in the implementation of software should be clearly defined.

d. There should be clearly defined and appropriate metrics to assess the dependency among components.

e. Dependencies generate complexity overheads in the overall software architecture . Extra care is required to implement and assess component dependencies.

1.7.7 Component Composition

Composition defines the capability of the component to be integrated with other compo-nents. Composition among components results in component-based software. The prop-erty of composability defines the usability of the component in the software. It is one of the fundamental properties of the component. Overall behavior of the software depends on the flexibility or rigidity of the components in respect of composability. Composition allows the component to behave according to the structure of the design. The composition of a component is its individual property. In component-based software development, components are assumed to be self-composable as well as composable independently.

Some basic issues regarding component composition are:

a. Composability state, that is, local or global, should be clearly defined.

b. Composing one component with other components should not affect the properties of those components.

c. No standard method or technique is available for composability of a component. Defining a methodology to fit all the components in the software is a big challenge.

d. Specific components require specific interfaces. In such cases, the number of inter-faces becomes an overhead in terms of cost and complexity.

e. Semantics of the components should be containable so that the process of composi-tion is easier.

Summary

This chapter is divided into two sections: the first focuses on the basics of software engi-neering; and the second is devoted to component-based software engineering.

The first section covers software engineering definitions used in academia and the industry. Various models suggested by researchers are discussed in this section. We have divided these models into two categories: traditional and advanced engineering models.

- Traditional models include:
  - Waterfall model
  - Incremental model
  - RAD model
  - Prototype model
  - Spiral model
- Advanced models include:
  - Agile model
  - Component-based model
  - Aspect-oriented model
  - Cleanroom development model

The second section of this chapter considers the basics of component-based software as well as current issues in this domain. The evolution of component-based software engineering is discussed in detail, with its advantages and disadvantages. This chapter also focuses on the features and attributes of the basic unit of this domain, that is, the component.

References

Agile Alliance. 2003 *Principles behind the Agile Manifesto*. http://agilemanifesto.org/principles.html. (Accessed on March 15, 2019.)

Atkinson, C. et al. 2002. *Component-Based Product-Line Engineering with UML*. Addison-Wesley, London.

Basili, V. R. and B. Boehm. 2001. "Cots-Based Systems Top 10 List." *IEEE Computer*, 34(5): 91–93.

Bauer, F. et al. 1968. *Software Engineering: A Report on a Conference Sponsored by NATO Science Committee*. NATO, Brussels.

Beck, K. et al. 2001. "Manifesto for Agile Software Development." www.agilemanifesto.org/. (Accessed on April 22, 2019.)

Bennatan, E. M. 2003. *Software Project Management: A Practitioner's Approach*. McGraw-Hill, New York.

Berssoff, E. H. and A. M. Davis. 1991. "Impacts of Life Cycle Models on Software Configuration Management." *Communications of the ACM*, 8(34): 104–118.

Boehm, B. W. 1981. *Software Engineering Economics*. Prentice Hall, Englewood Cliffs, NJ.

Boehm, B. W. 1986. "A Spiral Model for Software Development and Enhancement," *ACM Software Engineering Notes*, 14–24.

Boehm, B. W., M. Pendo, A. Pyster, E. D. Stuckle, and R. D. William. 1984. "An Environment for Improving Software Productivity." *IEEE Computer*, 17(6): 30–44.

Booch, G. 1987. *Software Components with Ada: Structures, Tools and Subsystems*. Benjamin-Cummings, Redwood, CA.

Bradac, M., D. Perry, and L. Votta. 1994. "Prototyping a Process Monitoring Experiment." *IEEE Transactions on Software Engineering*, 20(10): 774–784.

Brereton, B. and D. Budgen. 2000. "Component-Based Systems: A Classification of Issues." *IEEE Computer*, 33: 54–62.

Brown, A. W. 2000. *Large-Scale, Component-Based Development*. Prentice Hall, Upper Saddle River, NJ.

Brown, A. W. and K. C. Wallnau. 1998. "The Current State of CBSE." *IEEE Software*, 15(5): 37–46.

The Cambridge Dictionary. 2018. https://dictionary.cambridge.org/dictionary/english/component. (Accessed on June 4, 2018.)

Capretz, L. 2005. "Y: A New Component-Based Software Life Cycle Model." *Journal of Computer Science*, 1(1): 76–78.

Cockburn, A. 2002. *Agile Software Development*. Addison-Wesley, Boston, MA.

D'Souza, D. F. and A. C. Wills. 1998. *Objects, Components, and Frameworks with UML: The Catalysis Approach*. Addison-Wesley, Reading, MA.

Elrad, T., R. Filman, and A. Bader (eds.). 2001. "Aspect Oriented Programming." *Communications of ACM*, 44(10): 29–32.

Fowler, M. and J. Highsmith. 2001. "The Agile Manifesto." *Software Development Magazine*, www.sdmagazine.com/documents/s=844/sdm0108a/0108a.htm.

Gaedke, M. and J. Rehse. 2000. "Supporting Compositional Reuse in Component-Based Web Engineering." In *Proceedings of ACM Symposium on Applied Computing (SAC '00)*. ACM Press, New York, 927–933.

Gilb, T. 1988. *Principles of Software Engineering Management*. Addison-Wesley, Reading, MA.

Gill, N. S. and Balkishan. 2008. "Dependency and Interaction Oriented Complexity Metrics of Component Based Systems." *ACM SIGSOFT Software Engineering Notes*, 33(2): 1.

Heineman, G. and W. Councill. 2001. *Component-Based Software Engineering: Putting the Pieces Together*. Addison-Wesley, Boston, MA.

Hevner, A. R. and H. D. Mills. 1993. "Box Structure Methods for System Development with Objects." *IBM Systems Journal*, 31(2): 232–251.

Hopkins, J. 2000. "Component Primer." *Communications of the ACM*, 43(10): 28–30.

IEEE. 1987. *Software Engineering Standards*. IEEE Press, New York.

IEEE Standard 610.12-1990. 1993. *Glossary of Software Engineering Terminology*. IEEE, New York, ISBN: 1–55937–079–3.

Jacobson, I., M. Griss, and P. Jonsson. 1997. *Software Reuse: Architecture, Process and Organization for Business Success*. ACM Press, New York.

Jianguo, C., H. Wang, Y. Zhou, and D. S. Bruda. 2011. "Complexity Metrics for Component-Based Software Systems." *International Journal of Digital Content Technology and Its Applications*, 5(3): 235–244.

Jiangou, C., W. K. Yeap, and D. Braud. 2009. "A Review of Component Coupling Metrics for Component-Based Development." In *World Congress on Software Engineering*. IEEE Computer Society, Los Alamitos, CA, ISBN: 978–0–7695–3570.

Johnson, L. and D. R. Harris. 1991. "Sharing and Reuse of Requirements Knowledge." In *Proceedings of KBSE-91*. IEEE Press, Los Alamitos, CA, 57–66.

Kirti, T. and A. Sharma. 2012. "Reliability of Component Based Systems—A Critical Survey." *WSEAS Transactions on Computers*, 11(2): 45–54.

Kruchten, P. [1998] 2003. *The Rational Unified Process: An Introduction*, 3rd edn. Addison-Wesley, Upper Saddle River, NJ.

Krueger, C. W. 1992. "Software Reuse." *ACM Computing Surveys (CSUR)*, 2: 131–183.

Linger, R. 1994. "Cleanroom Process Model." *IEEE Software*, 11(2): 50–58.

Lin, M. and Y. Xu. 2003. "An Adaptive Dependability Model of Component-Based Software." *ACM SIGSOFT Software Engineering Notes*, 28(2): 10.

McIlroy, M. D. 1976. "Mass Produced Software Components." In J. M. Buxton, P. Naur, and B. Randell, eds., *Software Engineering Concepts and Techniques*, NATO Conference on Software Engineering. Van Nostrand Reinhold, New York, 88–98.

Maiden, N. and A. Sutcliffe. 1991. "Analogical Matching for Specification Reuse." In *Proceedings of KBSE-91*. IEEE Press, Los Alamitos, CA, 108–116.

Meyer, B. 2003. "The grand challenge of trusted components." In *Proceedings of IEEE ICSE*, Portland, OR, 660–667.

Nadeesha, G. R. 2001. History of Component Based Development. infoeng.ee.ic.ac.uk/~malikz/surprise2001/nr99e/article1. (Accessed on April 7, 2019.)

Object Management Group. 2000. "Unified Modeling Language Specification. Version 1.3." Retrieved from https://www.omg.org/spec/UML/1.3/PDF. (Accessed on March 20, 2019.)

Oxford Dictionary. 2018. https://www.oxfordlearnersdictionaries.com/definition/american_english/component. (Accessed on June 4, 2018.)

The Oxford English Dictionary. 2018. https://www.lexico.com/definition/component. (Accessed on June 4, 2018.)

Pressman, S. R. 2005. *Software Engineering: A Practitioner's Approach*, 6th edn. TMH International Edition, Boston, MA.

Royce, W. W. 1970. "Managing the Development of Large Software Systems: Concepts and Techniques." In *Proceedings of WESCON*, Los Angeles, CA.

Schach, S. 1990. *Software Engineering*. Vanderbilt University, Aksen Association.

Shepperd, M. 1988. "A Critique of Cyclomatic Complexity as Software Metric." *Software Engineering Journal*, 3(2): 30–36.

Sparling, M. 2000. "Lessons Learned—Through Six Years of Component-Based Development." *Communications of the ACM Journal*, 43(10): 47–53.

Szyperski, C. 1999. *Component Software—Beyond Object-Oriented Programming*. Addison-Weseley, Boston, MA.

Szyperski, C. and C. Pfister. 1997. "Component-Oriented Programming." In Mühlhäuser, ed., *WCOP'96 Workshop Report*, dPunkt.Verlag, 127–130.

Tiwari, U. K. and S. Kumar. 2014. "Cyclomatic Complexity Metric for Component Based Software." *ACM SIGSOFT Software Engineering Notes*, 39(1): 1–6.

Tiwari, U. K. and S. Kumar. 2016. "Components Integration-Effect Graph: A Black Box Testing and Test Case Generation Technique for Component-Based Software." *International Journal of Systems Assurance Engineering and Management*, 8(2): 393–407.

Vitharana, P. 2003. "Design Retrieval and Assembly in Component-Based Software Development." *Communications of the ACM*, 46(11): 97–102.

Webster's Dictionary. 2018. https://www.merriam-webster.com/dictionary/component. (Accessed on June 4, 2018.)

2

Component-Based Development Models

2.1 Introduction

This chapter discusses component-based software engineering development models proposed by eminent people from the fields of industry, research and academia. These models provide a broad understanding of the component-based development process. In component-based software development, huge and complex software is divided up into modules or components based on functionality, services or similar criteria. Almost every model focuses on components being reused and then deposited in the repository for future use. As we know, components and especially reusable components are the building blocks of component-based development. Reusable components are assembled and integrated to fulfill the requirements of the proposed software. Our focus here is on utilizing reusability rather than new development from scratch.

2.2 Y Life-Cycle Model

In his classic paper, "Y: A New Component-Based Software Life Cycle Model," Luiz Fernando Capretz (2005) discussed a new component-based model that supports the idea of development *with* reuse and development *for* reuse (Figure 2.1). The Y model is iterative and features forward and backward loops for feedbacks and increments. This model includes all the basic features of component-based software development, including reusability after development and component composition on the basis of reusability. The major phases involved in the Y model are: domain engineering, frameworking, assembly, archiving, system analysis, design, implementation, testing, deployment and maintenance.

- **Phase 1:** *Domain Engineering*—The first phase of the Y model, which starts with the commencement of the software, is domain engineering, conducted not only to configure essential requirements and constraints, but also to analyze the domain of the application. In this phase basic entities and their functions are identified and defined, reusability and other component features are sought, and an abstract model reflecting a real-world solution is depicted.
- **Phase 2:** *Frameworking*—In this phase the attempt is to define a generic architecture for the proposed software. Authors try to define the structure of the application and

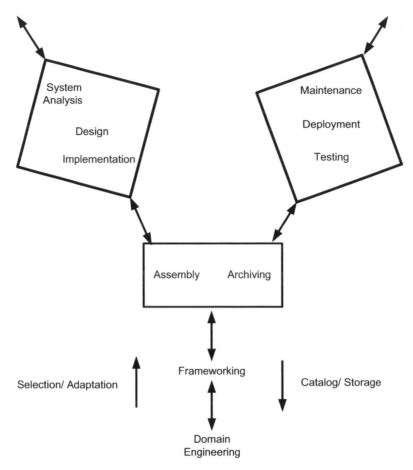

FIGURE 2.1
Y model for component-based software. (Adapted from Capretz 2005.)

also the generic structure for components so that they can be reused in more than one application. The frameworking phase attempts to:

a. Identify reusable components fit for the application.

b. Select candidate components.

c. Establish the semantic relationship among components.

d. Tailor components to fit into the framework of the software.

- **Phase 3:** *Assembly*—In this phase component selection or framework selection is made based on the requirements of the specific application's domain. Components are selected, based on their reusability, from archived components stored in the repository for reuse.

- **Phase 4:** *Archiving*—This is the process of submitting components to a repository for future use. This model supports the philosophy of reusability in two ways: exploring reusability during development and preserving reusability after development. Archiving involves the following activities:

a. Promoting reusability during development by using reusable components.

b. Depositing reusable components in the repository for future use.

c. Preserving components of a varied nature, as some components are generalized so that they can be placed easily, while others are specific.

d. Components should be properly documented and extensively cross-referenced so that they can be easily reused.

- **Phase 5:** *System Analysis*—This phase of development is used to identify and specify fundamental requirements as well as to define an abstract model of the application. In addition, this phase is used to:

  a. Create an abstract (physical or conceptual) representation of the system in the form of a model.

  b. Identify major components and modules of the application.

  c. Identify constraints of the application.

  d. Identify and configure basic features of the application.

- **Phase 6:** *Design*—This is the process of configuring solutions from the available resources. Application designers try to provide solutions easily and efficiently. The most suitable components are selected from the libraries, used and reused. Design is an iterative and repetitive process which takes time. This process should provide enough details for the next phase, implementation, to commence.

- **Phase 7:** *Implementation*—This is the phase when the developer has to decide whether the code for a particular feature or functionality is available to be reused or must be developed. In this phase, best-fit components are implemented and unavailable components are developed. If necessary, available components are modified and reused.

- **Phase 8:** *Testing*—Testing is done at two levels: component level and application level. Unit testing is applied to test individual components. When components are assembled, integration testing is applied. Interfaces are also tested, as they are used as a mechanism for the interaction of two components. After components are integrated and tested, various types of testing are applied and performed, including checks on the functioning of the proposed system.

- **Phase 9:** *Deployment*—On conclusion of the testing phase, the application is ready to install. Users are provided with proper documentation and training. The objective of this phase is not only to deploy the application at the user site but to make the user comfortable with the software. Ongoing help is provided to the user.

- **Phase 10:** *Maintenance*—This model argues that after deployment, software should be continuously maintained in order to keep pace with the changing environment. Components and component-based software are regularly updated according to the user's requirement. Updates compatible with older versions should be incorporated.

2.2.1 Key Findings

- This is one of the component-based development models that includes the reusability feature during as well as after development.

- It supports both features of development, that is, it defines activities in iterative as well as in incremental manner.

- This model is suitable for small-scale as well as large-scale applications, and supports both bottom-up and top-down approaches.

- Testing is done at two fundamental levels: at individual component level; and at the system as a unit level.

2.2.2 Critiques

- Although phases are defined and sub-activities are clearly mentioned, implementation details are not discussed and there is no mention of how these activities will take place.
- Overlapping of activities in different phases is quite common.
- Component selection criteria are not defined.

2.3 BRIDGE Model

Ardhendu Mandal (2009) proposed a component-based model named BRIDGE containing 13 phases (Figure 2.2). His model contains almost all the development phases that a model should contain. The BRIDGE model starts with requirements analysis, passing through feasibility and risk analysis, architecture design, detailed design, pattern and component search, coding, system building, validation, deployment and on-site testing, and ending with the maintenance phase. Detailed descriptions of every phase are provided.

- **Phase 1:** This is the initial phase of the model which focuses on three broad activities; requirement analysis, verification and specification.
  - *Requirement analysis*: This activity starts with the gathering of requirements from customers and uses standard requirement-gathering techniques. The requirements are analyzed to remove deficiencies such as redundancy, ambiguity and inconsistency.
  - *Verification*: After gathering and analyzing, the requirements are verified and the exact requirements configured.

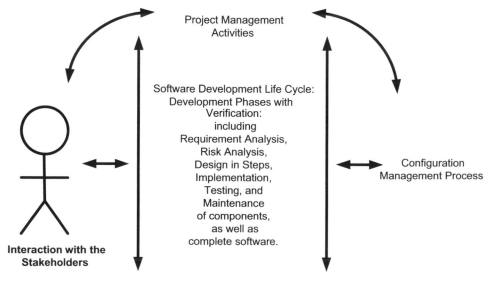

FIGURE 2.2
BRIDGE model for component-based software. (Adapted from Mandal 2009.)

- *Specification*: After verification, the requirements are well specified and documented in a software requirement specification that can be referenced in future. This specification document works as an intermediary between developer and customer.

- **Phase 2:** The focus of this phase is on proving the suitability of the application and considering alternate solutions. The four activities in this phase are feasibility analysis, risk analysis, verification and specification.

  - *Feasibility analysis*: Feasibility analysis covers all types of feasibility, such as economic, technical and operational feasibility. Systems analysts and the customer representatives jointly finalize the constraints and suitability of the proposed software.

  - *Risk analysis*: Risks that may occur during and after development are identified and recorded in the risk specification document.

  - *Verification*: Verification of the feasibility and risk analyses is carried out to ensure the optimal solution is achieved.

  - *Specification*: The final activity in this phase is the preparation of documents related to the issues identified and verified. Feasibility reports covering all types of feasibility and risk specification documents are generated for future reference.

- **Phase 3:** Phase 3, the beginning of the design phase, concentrates on designing the high-level, abstract architecture of the application. This phase is divided into three major activities: software architecture design (SAD), verification and specification. There are no implementation-related issues in this phase.

  - *SAD*: This is a high-level design architecture activity which focuses on identifying the sub-systems and/or components of the proposed system. Communication interfaces for these components and building blocks are also configured, so that an abstract architecture of the complete application can be prepared. This architecture design should contain all the necessary functional requirements as defined in the SRS document. Different stakeholders of the application are involved in this activity.

  - *Verification*: The design must now be verified to ensure it matches the requirements of the proposed system.

  - *Specification*: The output of this phase is the SAD document, which includes all the details of the abstract architecture design.

- **Phase 4:** In the BRIDGE model, the design phase is divided into two parts: abstract design and detailed design. Phase 4 covers the detailed design of the proposed system and includes the following activities:

  - *Detailed design*: Covers the low-level design including the implementation details of the application. The high-level or abstract design is translated into low-level implementation. Algorithms can be implemented, and data structures identified and defined in this phase.

  - *Verification*: The detailed design is verified, using the SRS document to identify any mismatch.

  - *Specification*: Finally the verified design is documented in the software design document, which is used as the basis for later phases.

- **Phase 5:** This phase covers the identification and selection of suitable components from the repository. Activities performed in this phase are:
  - *Component selection*: The goal of this activity is to select components according to the proposed system design. Specific components fulfilling the software requirements as well as generalized components providing more than one function are identified.
  - *Verification*: Verification is carried out after component selection.
  - *Specification*: Specifications are documented in the component specification document.
- **Phase 6:** This phase is devoted to coding and testing the proposed software. The major activities are:
  - *Coding*: For modules and functionalities for which proper components are not available, coding is performed using standard guidelines.
  - *Unit testing*: During coding, components are tested individually. Unit testing is the default testing performed on the components. These components must work according to their specifications.
  - *Component submission*: Newly developed components should be submitted to the repository so that they can be reused.
  - *Verification*: After testing, components are verified against the design documents. They must demonstrate compatible behavior as defined in the design documents.
  - *Specification*: The end of this phase produces testing documents in the form of unit-level test cases. Specifications of newly developed components are also included.
- **Phase 7:** This phase defines the development of the system as a unit from the individual sub-systems of previously development components and newly developed components or similar parts. This phase is time-consuming as it includes the following activities:
  - *System building*: Individual components are grouped together, according to the design, to form the proposed application.
  - *Component integration*: When different components are integrated, a number of issues may arise, which are resolved in this phase.
  - *Integration testing*: It may be that components working individually do not work well when they are integrated. Such problems should be dealt with in this phase.
  - *System testing*: When the complete system is integrated, the highest level of testing is performed to validate the behavior of the overall application. This type of testing is called system testing.
  - *Verification*: Verification is carried out on the basis of documents from previous phases.
  - *Specification*: Finally, documents are generated in the form of test plans, test designs and test cases.
- **Phase 8:** This phase provides activities related to system validation, that is, the compatibility of the proposed application with the functional requirements mentioned in

the SRS document. Quality factors are also measured in this phase. At the end of this phase a validation report is generated and verified.

- **Phase 9:** Phase 9 covers issues related to deployment and implementation of the developed system at the customer's site, including:
  - *Deployment and implementation*: Deliver the system to the client and implement it at their site. This installation and deployment may require help and support to be provided to the client. During development, necessary changes should be made in the application according to the customer's requirements.
  - *Documentation and training*: After the deployment of the application, clients and/ or users may need proper training and guidance. This activity emphasizes training and appropriate documentation for the client.

- **Phase 10:** The newly deployed system may generate new errors after installation, or environmental factors like the customer's hardware or other software may prove incompatible with the software. Provision should be made to deal with such problems. This phase focuses on on-site system testing to deal with such unforeseen events. The output of the phase is an on-site system testing report.

- **Phase 11:** This is the maintenance and user support phase, which starts after installation of the system at the client's site. Maintenance may include various types of activities that help the client after deployment. Maintenance reports are generated so that appropriate modifications can be performed at the client's wish.

- **Phase 12:** This phase, the configuration management phase, includes tools and techniques to manage and control modifications made to the proposed software in any phase. Changes suggested by the customer are documented and made, cost-effectively, in the next version. Documents generated during development of previous phases are used to keep track of the proposed changes.

- **Phase 13:** The final phase of this development model is the project management phase. Although it is mentioned last, it is in fact implemented from the beginning. Project planning, monitoring and similar activities are performed to keep control and manage the development process.

2.3.1 Key Findings

The major features of this development model are (Mandal 2009):

- It is a very flexible component-based model with customers involved in almost every phase of development.
- It works as a bridge between the end points of customer and project team, emphasizing communication between customer and management team in every phase.
- It includes verification activities in every phase of development.
- There is a major emphasis on design of the proposed software, with two phases: abstract design and detailed design.
- Risk analysis is included in development.
- Relevant documents are produced in every phase to facilitate the next phase or to use for future reference.

2.3.2 Critique

Mandal also provides the following critiques of his model:

- The model lacks implementation details. The development phases are mostly general theoretical concepts.
- Pattern matching and component selection are mentioned as important activities, but the matching criteria and component selection techniques are not discussed.
- Validation activity is mentioned but validation methods are not defined.
- This model appears to be both complex and time-consuming.

2.4 X Model

Gill and Tomar (2010) proposed a modified development process model for component-based software. This model supports the concept of reusability during and after development. The X model divides development into two broad approaches: composition based and generation based (Figure 2.3).

This model contains eight phases: domain engineering, domain analysis and specification, component and system architecture, design, coding and archiving, component testing, assembly and system testing.

- **Phase 1:** *Domain Engineering*—Like the Y model, this model starts with the domain engineering phase, devoted to analyzing the specific application's domain so that components can be identified, cataloged, developed or reused in the proposed application.

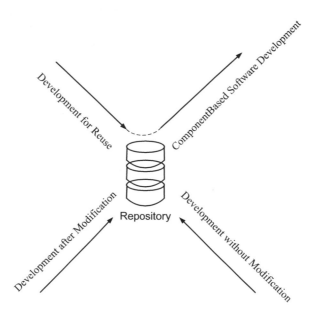

FIGURE 2.3
X model for component-based software. (Adapted from Gill and Tomar 2010.)

If components are identified early enough, they can be reused in a variety of ways in varied types of applications. Requirements of the system, constraints and other functionalities are also considered in this stage.

- **Phase 2:** *Domain analysis and specification*—Gathered requirements as well as the applications domain are analyzed to identify the general scope of the reusability of components. In parallel, specifications are generated to record the findings of these two phases.

- **Phase 3:** *Component and system architecture*—This phase includes two major activities: defining the architecture of components and defining the architecture of the overall application. The components' architecture is designed in such a way that they can be reused in many applications.

- **Phase 4:** *Design*—The design phase describes the implementation details of the component as well as the overall system. These details include descriptions of methods, functions, algorithms, data design, structure representation and similar activities. This phase also includes the design or selection of interfaces used for component interaction. The system design includes not only the selection of components but also the design of components which are not available and need to be developed.

- **Phase 5:** *Coding and archiving*—This phase includes the following activities:
  - Implementing components selected from the repository.
  - Making changes in the selected components according to the design requirements of the proposed application, if required.
  - Coding of components that are not available and that are designed by the development team as a fresh component.
  - Submission of newly developed components to the component repository for future reuse in other applications.

- **Phase 6:** *Component testing*—Each component is tested individually in this phase. Although components are tested by the developer during coding, at the end of component development they are tested in regard to the requirements of the customer or to match the requirement specification. All types of components are tested in this phase, whether picked from the repository, modified or newly developed.

- **Phase 7:** *Assembly*—The next activity is to assemble the components according to the predefined design. All the components providing various functionalities are integrated according to the system design. Well-defined interfaces are used to integrate these components. During the assembly phase all documents generated in previous phases are considered.

- **Phase 8:** *System Testing*—Integrated components are tested as they should behave according to the pre-defined architecture. When components are assembled, however, they may generate new and different types of problems. Components may perform well individually, but after integration they may create errors or faulty outputs. Components may be incompatible with each other or with the interfaces. Aspects of the customer's environment, such as hardware or other applications, may be incompatible with the proposed software. These types of issues are resolved in the final phase of the component-based development model.

2.4.1 Key Findings

- The X model supports reusability at three levels: architecture level, modular design level and framework level.
- This model defines use of black-box as well as white-box components, both with and without modifications.
- This component-based development model follows the reusability approach during as well as after development.
- This model is also suitable for small-scale as well as large-scale applications, as well as for both bottom-up and top-down approaches.
- Testing is done at two fundamental levels, individual component level and system as a unit level.

2.4.2 Critiques

- Implementation details are not discussed. Though phases are defined and sub-activities are clearly mentioned, there is no discussion of how these activities will take place.
- Selection criteria for components are not defined.
- No risk analysis activity is defined.

2.5 Umbrella Model

In 2011, Dixit and Saxena (2011) proposed a component-based model named the Umbrella model (Figure 2.4). The phases of the Umbrella model are similar to those of other proposed component-based models, as shown in Figure 2.4.

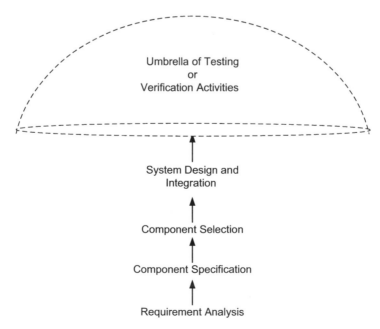

FIGURE 2.4
Umbrella model for component-based software. (Adapted from Dixit and Saxena 2011.)

The Umbrella model divides software development into three steps: design, integration, and deployment or run-time.

- **Design phase:** The design phase includes:
  - Suitable and proper component selection from the repository, that is, components that fit into the design are sought and selected for development.
  - Components that are not available in the repository must be defined and constructed or developed by the developer.
- **Integration phase:** The objective of this phase is to integrate components according to the design. Selected components and developed components are put together according to the pre-defined architecture.
- **Deployment or run-time phase:** This phase is devoted to the deployment and installation of the application, including component libraries and other essential parts of the system, to execute the proposed software.

The formal steps of the Umbrella model are:

Requirement specification: Development starts with the process of specifying requirements for the proposed project. According to this model, in this phase it is the customer and developer who are involved in this activity.

Analysis of components: Components are analyzed, defined and chosen from the repository as per the customer's requirements. It is the developer's job to analyze the component.

Modifications to the requirements: Requirements may be changed or modified to select the most suitable component.

System design with reuse: This phase emphasizes reusability during the design of the proposed application.

Development and integration: Components are integrated and deployed according to the predefined architecture.

System validation: This is the broad phase that is normally applicable to all phases of the development. During different phases of system validation, the different testing activities included in the Umbrella model are requirement testing, specification testing, selection testing and integration testing.

2.5.1 Key Findings

- Different types of testing activities are involved during different phases of development.
- Components are selected iteratively, that is, components and requirements are matched accordingly in phases.
- This model includes verification activity in every phase of development.

2.5.2 Critique

- This model lacks implementation detail. Most of the development phases are general theoretical concepts.
- Component selection is mentioned as an important activity, but criteria for matching and techniques of component selection are not discussed.
- Risks associated with developing software or components are not discussed at all. There is no risk analysis phase in the model.

2.6 Knot Model

Chhillar and Kajla (2011) suggested a development model for component-based software that emphasizes risk analysis in addition to reusability, modularity and feedback in every phase (Figure 2.5). The model includes four phases: reusable component pool, new component development, modification of existing components and development of component-based software. When components are developed, they are submitted to the component pool for future reuse. Components can be reused with or without modifications as per the requirements of the software.

- **Reusable component pool phase:** All the components are stored and managed in the component repository. When components are developed, they are submitted to the pool for reuse in other applications. This pool contains all types of components, including modified and newly developed.

- **New component development phase:** When requirements are not fulfilled by existing or modified components, new developments is carried out. We can use any traditional model to develop these components. New component development starts with cost estimation and risk analysis, and is followed by design and coding activities and software analysis. Component testing and feedback commences the development of new components.

- **Modification of existing components phase:** Existing components can be modified when they fail to provide the desired features. Components are selected from the repository, their adaptability is checked and risk analysis is carried out. Coding is then performed according to the requirements, and finally the component is tested. After modification the new version of the component is submitted to the pool so that it can be reused.

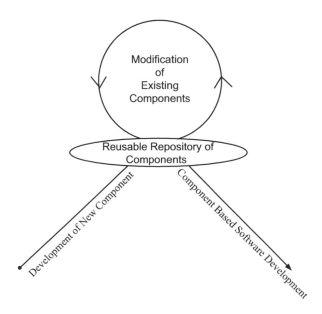

FIGURE 2.5
The Knot model for component-based software. (Adapted from Chhillar and Kajla 2011.)

- **Development of component-based software:** This phase includes cost estimation, risk analysis, assembly of components, system testing, implementation, feedback and finally maintenance of the software.

2.6.1 Key Findings

- This model can be applied to variable scales of software as well as component development.
- The Knot model includes risk assessment in every phase of development, which reduces the risk of unforeseen events in later phases of development.
- This model helps to enhance the reusability of components as well as reusability of component-based software.

2.6.2 Critiques

- The component pool may become too large to manage if components and developed software products are stored after development.
- The Knot model does not suggest any method for storing or pooling of components or software products in the repository.
- No criteria for component selection are included in the model.

2.7 Elite Model

Nautiyal et al. (2012) proposed an Elite model for development of component-based software (Figure 2.6). The Elite model focuses on reusability of components during development and parallel development of components in software. It uses the Rapid Application Development model of component development. The Elite model consists of nine phases: requirement, scope elaboration, reusing existing components (without modifications),

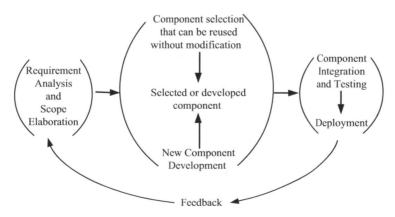

FIGURE 2.6
The Elite model for component-based software. (Adapted from Nautiyal et al. 2012.)

reusing existing components (with modifications), developing new components, integration of components, component testing, release and customer evaluation.

- **Requirement:** The Elite model starts with the requirement-gathering and analysis phase. Requirements are identified for the existing application or system domain. "The Requirement phase involves carrying out enough business/application/system modeling to define a meaningful build scope. A build delivers a well-defined set of business functionalities that end-users can use to do real work." (Nautiyal et al. 2012). The major objectives of this phase are:
  - Identifying the problem domain, with the emphasis on understanding the domain as well as the problem areas.
  - Categorizing and prioritizing requirements.
  - Setting the aims and objectives of the proposed software, considering previously defined priorities.
  - Identifying the boundaries of the proposed software, considering the limitations of available resources.
- **Scope elaboration:** This phase "emphasises on determining the illustrations of build requirements in the form of technical, behavioural and structural specifications." (Nautiyal et al. 2012). Requirements are prioritized, and three types of specifications are generated: technical specifications including resources required, together with their limitations; functionalities required by the proposed software; and architectural specifications of the proposed system.
- **Reusing existing components (without modifications):** After building the specifications, the next step is to identify components which can be reused as they are, without any modifications. Interfaces are identified and minor modifications are carried out as required.
- **Reusing existing components (with modifications):** It is possible that some components may be reused with minor or major modifications. But these modifications must conform to the scope of the proposed application. This step is used to identify such modifiable components that can later be reused.
- **Developing new components:** Some components may be required to be developed from scratch. These components should have defined interfaces and must be designed and implemented according to the specifications.
- **Integration of components:** When all components are identified that fit the design of the software, they must be integrated and tested. Components are integrated in a bottom-up approach, and then clusters of components are aerated. These clusters are then integrated to develop the software.
- **Component testing:** Component testing emphasizes testing individual as well as integrated components. For testing we can use black-box and white-box testing methods.
- **Release:** The next phase is the deployment of the system, including fixing bugs encountered during deployment. Release includes user-oriented documentation and training.
- **Customer evaluation:** When the software is released and deployed to the customer's site, the customer is free to evaluate it. Evaluation is the process of taking and giving feedback from/to the customer. Necessary changes are made to satisfy the customer.

2.7.1 Key Findings

- The Elite model focuses on reusability of components.
- This model includes almost all the phases of development.
- Structural, behavioral and technical specifications are useful for designers, coders and testers during the development of the component as well as during the development of the software.
- Feedback is collected from customers and applied to the next release after re-development of the component or the proposed software.

2.7.2 Critiquess

- No actual implementation is available for this model. All the phases and methods are theoretical.
- Testing methodologies and techniques are not defined.
- There is no risk analysis for any part or component of the proposed software.
- Integration issues are not covered at all.

Summary

This chapter covers development models suggested by eminent researchers in the context of the component-based software environment. Development models are critically discussed so that their advantages and disadvantages can be clearly determined. Every model is suitable for specific scenarios. The major focus of every model is on domain engineering, component repository, development with reuse and reusability after development, and these development phases are commonly covered in each model. Our discussion includes models which are specific to component-based software development.

References

Capretz, L. F. 2005. "Y: A New Component-Based Software Life Cycle Model." *Journal of Computer Science*, 1(1): 76–82, ISSN: 1549-3636.

Chhillar, R. S. and P. Kajla. 2011. "A New-Knot Model for Component Based Software Development." *International Journal of Computer Science Issues*, 8(3), 480–484. ISSN: 1694–0814.

Dixit, A. and P. C. Saxena. 2011. "Umbrella: A New Component-Based Software Development Model." *International Conference on Computer Engineering and Applications, IPCSIT*, Vol. 2. IACSIT Press, Singapore.

Gill, N. S. and P. Tomar. 2010. "Modified Development Process of Component-Based Software Engineering." *ACM SIGSOFT Software Engineering Notes*, 35(2): 1–6.

Mandal, A. 2009. "BRIDGE: A Model for Modern Software Development Process to Cater the Present Software Crisis." *IEEE International Advance Computing Conference (IACC 2009)*, Patiala, India, 6–7.

Nautiyal, L. et al. 2012. "Elite: A New Component-Based Software Development Model." *International Journal of Computer Technology & Applications*, 3(1): 119-124, ISSN: 2229–6093.

3

Major Issues in Component-Based Software Engineering

3.1 Introduction

Component-based software engineering addresses the expectations and requirements of customers and users, just as other branches of software engineering do. It follows the same development steps and phases as other development paradigms. Standard software engineering principles apply to applications developed through component-based software engineering. Reusability gives the development team the opportunity to concentrate on the quality aspects of the software (Naur and Randall 1969). The principle of reusability is applied to development not only of the whole system but also of individual components. Development *with* reuse focuses on the identification, selection and composition of reusable components. Development *for* reuse is concerned with the development of such components as may be used and then reused in many applications, in similar and heterogeneous contexts.

This chapter discusses a number of issues in the context of component-based software development: *reuse and reusability of components, integration and integration complexities, issues related to testing, reliability issues* and *quality issues*.

3.2 Reuse and Reusability Issues

Reusability is the focal point of component-based software development. Software reusability is defined as the effective reuse of pre-designed and tested parts of experienced software in new applications. In CBSE, we integrate components of all classes according to the design architecture and application requirements. Components for which code is not available are called black-box components; those for which code and documentation are available are known as white-box components. In the literature, researchers have classified software reusability quantification methods into various categories.

Reusability can be broadly defined at three levels: statement level, function level and component level.

- *Statement-level* reusability relates to development of programs in common programming languages. Predefined keywords and syntax styles are reused in programs.

- *Function-level* reusability applies to predefined programming functions used by programmers and developers. These functions are either provided by the language library or defined by the user.

- *Component-level* reusability relates to development of large and complex software. Pre-developed components are reused according to the requirements of the software.

The following issues can be identified in the context of reusability:

- **Measures and metrics to assess reusability at individual component level:** Measures identified and defined in the literature are for conventional software modules, functions or object-oriented classes. Relatively less attention has been paid by researchers and practitioners to exploring the reusability of new, fully qualified, partly qualified and off-the-shelf components. In addition, a single type of reusability metric (black-box or white-box) is not suitable for all CBSE applications, since these complex applications contain both black-box and white-box components.

- **Measures and metrics to assess reusability at system level in component-based software engineering:** Reusability measures and metrics for component-based systems need to be defined. It is a broad area of research as universal measures to assess system-level reusability are lacking.

- **Factors affecting reusability of components:** While researchers have defined some factors affecting reusability, these factors are general. We need to identify factors that affect the reusability of components specifically in the component-based environment.

- **Factors affecting the reusability of component-based software:** Similarly, we need to identify factors affecting the reusability of the complete system. As the reusability of the system increases, there will be a corresponding reduction in both cost and time, and development effort.

- **Applications of reusability in component-based software:** Assessing the reusability of components or component-based software is not sufficient. We need to identify and define reusability applications in various areas of development as well as after development.

- **Reusability cost:** The cost of reusability is a comparatively less explored area. There is a lot of scope for identifying and defining the reusability cost of individual components as well as the component-based software.

- **Reusability effort:** This is the measure that computes the effort invested in the process of component reusability and component-based software. Reusability may require changes and modifications to components or interfaces.

- **Amount/level of reusability:** Some components may be reused as they are, while others may need a level of modification. The amount of reusability indicates the level of usability of the particular component in a heterogeneous environment. The amount of reusability needs to be assessed and stored as meta-information with the component.

- **Trade-off between adaptability and reusability:** Component adaptability directs the reusable behavior of the component. The more flexible and supportive the adaptability of the component, the more its reusability improves. The trade-off between these two aspects needs to be evaluated and properly defined.

- **Frequency of reusability:** Component usability in development increases as the choice and frequency of reuse of components increases. Frequency of reuse is an indirect measure of component quality.

- **Measures and metrics to assess the effects of reusability:** It is a common assumption that reusability has positive effects. Assessing the effects of reusability helps designers and developers to better understand the development of component-based software. Selection of inappropriate components may lead to negative effects during reuse.

- **Risks associated with reusability:** There are no standard techniques for risk analysis in the context of reusability in component-based software. However, there is great scope for estimating risks associated with component use or reuse, as well as overall risk factors associated with component-based software.

- **Selection methods for components other than for requirement criteria:** If we have a number of components available for similar purposes and we have to single out one of them, there should be some selection and verification criteria in addition to the requirement specification.

According to Poulin (1996) and Prieto-Diaz and Freeman (1987), reusability assessment techniques can be divided into two approaches, empirical and qualitative. Empirical approaches rely on objective data to define reusability attributes and characteristics, whereas qualitative approaches use qualitative software assessment techniques, including identifying and defining subjective guidelines and standards.

Prieto-Diaz and Freeman (1987) define attributes of a program and related metrics to compute reusability. They propose that reuse depends on size, program structure, documentation, programming language and reuse experience. Lines of code are used to count the size and cyclomatic complexity of a structure, rated from 0 to 10 for documentation; inter-module language dependency is used to estimate the difficulty of modification and experience of using the same module. Two levels of reuse are identified: reuse of idea and knowledge, and reuse of artefacts and components. The authors define a reusability model using the approach "to provide an environment that helps locate components, and that estimates the adaptation and conversion effort based on the evaluation of their suitability for reuse." (Prieto-Diaz and Freeman 1987).

The evaluation scheme is based on the following assumptions:

- The component collection is very large.
- It is possible that components providing similar functionalities are available.
- There are various levels of reusability.
- Different types of attribute are possible for a program.
- Developers need assistance to select the most suitable component.

Caldiera and Basili (1991) proposed one of the earliest methods of identifying and qualifying reusable components. They define cost, usability and quality as the three factors affecting reusability. Measures and metrics are used to identify qualifying components so that reusability can be identified and the exact component extracted. Component reusability is characterized using four metrics: (i) volume of operands and operators, using Halstead Software Science Indicators; (ii) cyclomatic complexity using McCabe's method; (iii) regularity, which measures the component's implementation economy; and (iv) reuse frequency, the indirect measure of its functional usefulness. The method also defines an organization reuse framework in which two organizations are identified: a "project organization" and an "experience factory."

The project organization constructs projects using and reusing currently developed and previously developed components. The experience factory is an organization that builds and packages the component. The activities of the experience factory are of two types: synchronous and asynchronous. The basic idea behind both activities is to enhance reusability through or after development.

The extraction process of components from the repository is also defined. Components to be extracted must first be identified in a three-phase process:

1. Reusability requirements
2. Component extraction
3. Application

The qualification process follows identification and consists of six steps:

1. Defining functional specification of component
2. Test-case construction
3. Component classification
4. User manual
5. Storage
6. Feedback

Caldiera and Basili (1991) define a range of reusability factors associated with a varied range of values to measure these attributes. These reusability factors are:

- **Cost:** This defines the cost of reusability. It includes the cost of searching for, selecting and adapting components.
- **Usefulness:** It includes common features of component behavior represented in other applications as well as various functionalities provided by the component.
- **Quality:** This defines the basic set of attributes directly or indirectly related to the component. It includes correctness, readability, testability, ease of modification and performance.

Barnard (1998) reused components developed in C++, Java and Eiffel to show his experimental findings. His work was based on estimations of the component's attributes, such as simplicity, genericity and understandability, through available properties, methods and defined interfaces. Barnard's work lists reusability factors in the context of traditional, object-oriented and component-based software, as:

- **Encapsulation:** This defines the encapsulation of data and associated methods. As the level of encapsulation increases, the degree of reusability also increases.

- **Cohesion:** Cohesion defines the binding level of sub-components within the particular component. Better cohesion increases the reusability of the component.
- **Coupling:** Defines the requirement for a particular component to perform its task, that is, level of interdependency. Reusability increases as coupling decreases.
- **Complexity:** Defines ease of use of the component. As complexity decreases, the level of reusability increases.
- **Instance and class variables:** These are the variables used by the methods, classes or components. Fewer variables means an increased level of reusability.
- **Depth of inheritance:** Depth is defined in terms of reusability. As the depth decreases, reusability increases.
- **Number of children:** As the number of children of a component or class increases, reusability also increases.
- **Correctness:** Correct components are more reusable. Components should possess the property of correctness in a defined environment.
- **Program size:** Size should be justifiable. As the size of the component increases, it becomes less usable.
- **Program documentation:** All types of documentation should be prepared for the component. Documentation increases the understandability of the component. Proper documentation increases the level of reusability.
- **Functions:** Defines the number of functions provided by the method. For better reusability it should be minimal, that is, one.
- **Complexity:** relates to the method's interface.
- **Robustness:** This is the property that defines techniques for handling exceptions by the component.
- **Portability:** Defines the level of ease with which a class or component can be reused in different environments.

Boxall and Araban (2004) find that the reuse level of a component is greatly affected by its understandability regarding the interfaces it uses. They derive the value of understandability using the attributes of the component's interfaces. Boxall and Arban consider interface size, counts of the argument, number of repetitions, scale of the repetitions and similar attributes to suggest some metrics for understandability. They use interfaces used for 12 components to provide data for their metrics. Tools used in their metrics are tested for component interfaces that are developed in C and C++ only. Metrics defined include *interface size*, the ratio of argument count and procedure counts, where argument count is the number of arguments included in public procedures and procedure count is the number of procedures declared publicly by an interface. Other metrics are *distinct argument count*, *argument repetition scale, identifier length* and *mean string commonality*. Their study is based on black-box components whose source code is not available. In cases where not much information is available, interfaces play a vital role in enhancing understandability of components.

Washizaki et al. (2003) suggested a reusability model for efficient reuse of black-box components. Factors affecting reusability include functions offered, and adaptability to new environments and varied requirements. The authors propose metrics for better understandability, adaptability level and portability features of components. Since these are black-box components, their code is unavailable. The authors use only available static information to define their set of metrics. The criteria in their metric model are:

- **Existence of meta-information:** Information about the features of the component helps to understand its usability.
- **Observability:** "Observability measures how easy it is to observe a component in terms of its operational behaviours, input parameters, and outputs." (Washizaki et al. 2003).
- **Customizability:** Measures the level of changes and modifications supported by the component.
- **External dependency:** Indicates the level of interdependency among components.

The authors also define a set of metrics for the existence of meta-information, rate of component observability, rate of component customizability, self-completeness of the component's return value and self-completeness of the component's parameters.

Bhattacharya and Perry (2005) focus on the integration contexts of components rather than just their internal attributes. They propose reusability estimations considering the integration architecture of the software, as well as component characteristics and architecture compliance metrics to measure issues related to particular properties and integration issues of components. They suggest "the use of software architecture descriptions as the context of a software component to use and reuse." The basic three properties of architectural descriptions are given as:

1. Descriptions of services provided by the interfaces
2. Descriptions of attributes
3. Descriptions of behavior.

These authors also propose a set of quantitative metrics for architecture descriptions, including "architecture compliance metrics and component characteristics metrics." A set of sub-metrics is also included. Coefficients used to define these metrics include input and output data, events related to inputs and outputs, pre-conditions and post-condition compliance coefficients. These are generic metrics applicable to all categories of components.

Gui and Scott's (2007) study shows that no efficient metrics are available to predict the effort of modification. They used Java components to suggest metrics to estimate the cohesion properties of intra-components and the coupling properties among them. The authors claim that these metrics are efficient predictors of the time, effort and amount of changes required to make a component more useful. Rank correlation and linear regression are used to evaluate the relative performances of proposed metrics.

Gui and Scott (2008) further define a suite of metrics to compute the reusability of Java components, and rank them according to their reusability. These metrics are used to estimate indirect coupling among components as well as to assess the degree of coupling.

Hristov et al. (2012) categorize reusability metrics into two broad classes: white-box metrics and the black-box metrics. *White-box reusability metrics* are based on the logic and coding structure of the software. *Black-box reusability metrics* are based on interfaces and other attributes, as code is not available for black-box components. The authors define factors directly affecting the reusability of component as:

- **Availability:** The measure that defines the speed of retrieval of the component. If availability is high then retrieval is easy.
- **Documentation:** Documentation helps to understand the component.
- **Complexity:** Complexity determines the ease of usability of the component. It also defines the level of ease of using the component in the new context.
- **Quality:** Quality is the measure of usability of the component in the defined context. Quality is directly related to the extent of fulfilment of requirements by the component as well as error-freeness and bug-freeness of the component.
- **Maintainability:** This relates to reusability after development and deployment of the component.
- **Adaptability:** The level of modification required is the level of adaptability of the component. It defines the ease with which the component can be reused in the new system.
- **Reuse:** It identifies the frequency of reuse. This is the actual reuse of the component in a similar or dissimilar environment.
- **Price:** The price of reusing a component also affects its reusability.

Poulin (1994) identifies two basic types of reusability model: empirical and qualitative. *Empirical models* use experimental data to estimate complexity, size, reliability and similar issues, which can be used by automated tools to estimate reusability. *Qualitative models* stay with predefined assumptions and guidelines to address issues like quality and certification.

Chen et al. (Chen and Lee 1993) present findings based on a large number of reused components. They computed size, program volume, program level, difficulty of developing and effort for all these reused components, concluding that to increase productivity we should decrease the values of these metrics.

Hislop (1993) defines the theory of function, form and similarity to compute software reusability. *Function* defines the actions of a component, *form* characterizes attributes like structure and size, and *similarity* identifies the common properties of components. The author uses well-defined metrics like McCabe's complexity metric in his calculation.

Wijayasiriwardhane and Lai (2010) suggest a size-measurement technique named "component point" to estimate the size of overall component-based systems. These component points can be reused as a metric to analyze components for future use. Three classes of component are defined according to their usability.

Lee and Chang (2000) suggest metrics including complexity and modularity of components to predict the reusability and maintainability of object-oriented applications. They define complexity metrics as "internal–external class complexity," and modularity metrics as "class cohesion coupling."

Cho et al. (2001) propose component reusability measures as the fraction of total interface methods and the number of interface methods in the component that have common functionality in their domain. Reusability is assumed to rise as the value of the ratio increases. They also define customizability metrics as the ratio between the customization methods and the method count present in the interface of that component.

Reusability measures defined in the literature can be analyzed by considering development paradigms such as conventional software and programs, object-oriented software, and component-based software. Table 3.1 provides a summary.

TABLE 3.1

Summary of Reuse and Reusability Issues

Paradigm	Measures and Metrics Used	Key Findings	Factors Affecting Reusability	Author(s)/ References
Conventional software and program	• Lines of code • Cyclomatic complexity • Rating from 0 to 10 for documentation • Inter-module language dependency to estimate difficulty of modification • Experience of using same module	• Defines some attributes of a program and related metrics to compute reusability • Proposes that reuse depends on size, program structure, documentation, programming language and reuse experience	• Size • Program structure • Documentation • Programming language • Reuse experience	Prieto-Diaz and Freeman (1987)
Component-based software	• Halstead Software Science Indicator to find volume • *Cyclomatic complexity* using McCabe's method.	• *Regularity* measures the component's implementation economy • *Reuse frequency*, the indirect measure of the functional usefulness	• Cost • Usability • Quality	Caldiera and Basili (1991)
Component-based software	• Automated tools to estimate reusability • Predefined assumptions	• Two categories of reusability: • *Empirical models* use experimental data to estimate complexity, size, reliability and similar issues • *Qualitative models* stay with pre-defined assumptions and guidelines to address issues like quality and certification	• Cost • Reliability	Poulin (1994)
Object-oriented software	• Uses reused components • Modularity • Complexity	• Findings based on a large number of reused components • To increase productivity we should decrease the values of these metrics	• Size • Volume • Level • Difficulty to develop • Effort	Chen and Lee (1993)
Component-based software	• Structure of components • McCabe's complexity	Defines reusability on the basis of: • *Function* defines the actions of a component • *Form* characterizes the attributes like structure and size • *Similarity* identifies the common properties of components	• Functions of the components • Size • Complexity of components	Hislop (1993)

Component-based software	• Reusable components • Automated tools	• Importance of reusability proposed in experimental findings • Work based on estimations of such component attributes as simplicity, genericity and understandability	• Available properties • Methods • Interfaces	Barnard (1998)
Object-oriented software	• Internal–external class complexity • Modularity metrics (class cohesion and class coupling)	To estimate the reusability and maintainability of object-oriented software metrics considering the complexity and modularity of components.	• Complexity • Modularity	Lee and Chang (2000)
Component-based software	• Metrics for reusability • Metrics for customizability	Proposed metrics for reusability and customizability: • The ratio of total number of interface methods to the number of interface methods in the component that have common functionality in their domain • Reusability is assumed higher as the value of ratio increases • Ratio between customization methods and total number of methods present in the interface of the component	• Total number of interface methods • Total number of common-function interface methods • Total number of customization methods • Total number of common-function customization methods	Cho et al. (2001)
Component-based software	• Size of interfaces • Number of arguments • Scale of repetition of arguments	• Study argues that reuse level of a component is greatly affected by component's understandability • Value of understandability of a component derived using its interface attributes	• Understandability • Interface property	Boxall and Araban (2004)
Component-based software	• Indicates these values as high or low • Structured relationships between understandability, adaptability and portability	• Component reusability model • Component's observability: readability of component/total properties • Component's customizability: writability of component/total properties	• Functionality of the component • Adapting to changes in requirements • Porting to new environment	Washizaki et al. (2003)

(continued)

TABLE 3.1 (*Continued*)

Summary of Reuse and Reusability Issues

Paradigm	Measures and Metrics Used	Key Findings	Factors Affecting Reusability	Author(s)/References
Component-based software	• Interaction contexts • Properties and attributes of components • CBS architecture	• Focused on integration contexts of components rather than just their internal attributes • Reusability estimations proposed considering integration architecture of the software • Component characteristics and architecture compliance metrics proposed to measure issues related to particular properties and integration issues of components	• Integration architecture	Bhattacharya and Perry (2005)
Object-oriented software	• Coupling and cohesion metrics	• Estimate coupling properties among inter-components and cohesion properties of intra-components	• Complexity of components	Gui and Scott (2007)
Component-based software engineering	• Coupling between objects • Response for class • Coupling factors	• Components ranked according to their reusability • Metrics to estimate indirect coupling, degree of coupling and functional complexity	• Number of changes made to the code • Time required to carry them	Gui and Scott (2008)
Component-based software	• Interface complexity • Interaction complexity • Internal logical files • External interface files • Number of operations and interactions	• Size-measurement metric named "component point" to estimate size of overall component-based system • Three classes of components: user, service and domain	• Size • Weighting factor low, average and high • Unadjusted component point	Wijayasiriwardhane and Lai (2010)
Component-based software	• Reusability metrics for available code • Interface properties	• *White-box reusability metrics* based on software internal logic and coding structure • *Black-box reusability metrics* based on interfaces and other attributes, as code is not available	• Internal code • Interfaces • Interaction • Complexity	Hristov et al. (2012)

3.3 Interaction and Integration Issues

Major complexities of component interaction and integration are:

- **Interaction and integration measures and metrics to capture the complexities for components in component-based software:** When components interact, they have different configurations, diverse complexity levels, coupling issues and incompatible interface problems which significantly impact the quality of a component-based system. It is not only sufficient but necessary to have appropriate metrics to quantify the complexities produced by these interactions and integrations. Most of the interaction, integration and complexity quantification methods available in the literature are program oriented, i.e., suitable for small-scale software. But for large, complex component-based software systems these metrics are inefficient.

- **Level of interactions and integration of components in component-based software:** Component-based software integrates different components providing different functionalities. Components interact with each other. We need measures and metrics to assess their level of interaction and integration.

- **Number of interactions made by individual components:** It is important to evaluate the number of interactions made by individual components in the system, as these ultimately affect the complexity of the software. The goal is to keep the number of interactions to a minimum.

- **Component integration effort:** Resources (hardware, software or human) required by the component to provide services should be identified.

- **Cost of integration of components:** It is quite common for components to be provided by or purchased from a third party. A particular component's integration cost contributes to the overall cost of the system. Integration cost estimation techniques are required to manage the trade-off between usefulness and cost of the component.

- **Effects of integration of components:** It is not sufficient merely to integrate components: the effects of integration must also be evaluated to assess the behavioral aspects of the component. Proper assessment methods are required to record the effects of integration.

- **Total time required for integration:** Component integration time is a major part of overall component-based software development. Measures to predict component integration time are essential.

Component-based software applications are composed of independently deployable components. These components are assembled in order to contribute their functionalities to the system. Technically this assembly is referred to as integration of and interaction among components. We have sufficient measures and metrics to assess the complexity of standalone programs as well as small-sized conventional software proposed and practiced by numerous practitioners. In the literature, the complexity of programs and software is treated as a "multidimensional construct" (Kumari and Bhasin 2011, Wake and Henry 1988). Complexity depends upon various factors in a program (Basili and Hutchens 1983).

McCabe (1976) defined a complexity measurement method based on the interactions among the statements of a program. He offered and implemented graph-theoretic notions in programming applications, using a code control-flow graph to compute the complexity.

In a control-flow graph, a sequential block of code or a single statement is represented as a node, and control flows among these nodes are represented as edges. The cyclomatic complexity metric is easy to compute and maintain, as well as giving the relative complexity of various designs. This method is applicable both to standalone programs and to hierarchical nests of subprograms.

Paradigm:

Conventional program/hierarchical software

Method:

McCabe used a set of programs developed in FORTRAN to illustrate his implementations, with graph-theoretic notation to draw the control-flow graph, where a graph denoted G has n number of nodes, e number of connecting edges and p number of components.

Cyclomatic complexity V(G) is calculated as

$$V(G) = e - n + 2p,$$

where 2 is the "result of adding an extra edge from the exit node to the entry node of each component module graph" (Pressman 2005). In a structured program where we have predicate nodes, complexity is defined as

$$V(G) = \text{number of predicate nodes} + 1,$$

where predicate nodes are the nodes having two and only two outgoing edges.

In his implementations, McCabe defined a cyclomatic complexity value of a program of less than 10 as reasonable. If a program has a hierarchical structure, that is, one subprogram is calling other one, the cyclomatic complexity is the summation of individual complexities of these two subprograms and is given as

$$V(G) = v(P_1 + P_2) = v(P_1) + v(P_2),$$

where P_1 and P_2 are two subprograms and P_1 is calling P_2.

3.3.1 Key Findings

- Complexity depends not on the size but on the coding structure of the program.
- If a program has only one statement then it has complexity 1. That is, $V(G) \geq 1$.
- Cyclomatic complexity V(G) actually defines the number of independent logics/paths in the program.

3.3.2 Metrics Used

- Lines of code
- Control flow of statements
- Interaction among statements
- Independent paths from source to destination
- Vertices and edges

3.3.3 Factors Affecting Interaction and Integration Complexity

- Structure of the program
- Forward and backward loops
- Branching statements
- Switch cases in the program

3.3.4 Critique

- The same program written in different languages or with different coding styles or structures may have different complexities.
- Intra-module complexity of simple structured programs can be achieved easily, but for inter-module complexity, this metric produces a misleading output.

Halstead (1977) identified a complete set of metrics to measure the complexity of a program considering various factors. He used recognized scientific theories to prove his study and metrics on complex production of software. These metrics include program vocabulary, length, volume, potential volume and program level. Halstead proposed methods to compute the total time and effort to develop the software. These metrics are based on the lines of codes of the program. Halstead also defined the relationship between these factors and metrics of programs.

Paradigm:

Conventional program/hierarchical software

Method:

Halstead proposed software science to examine the algorithms developed in ALGOL and FORTRAN. Halstead considered the algorithms/programs as a collection of "tokens," that is, operators and operands. He defined program vocabulary as the count of distinct operators and distinct operands used in the program. The count of total operators and operands used in a program is proposed as the program length. The program volume is defined as the storage volume required to represent the program, and the representation of the program in the shortest way without repeating operators and operands is known as potential volume.

$$\text{Program vocabulary: } n = n_1 + n_2,$$

where n_1 and n_2 are the count of unique operators and operands respectively.

$$\text{Program length } N = N_1 + N_2,$$

where N_1 and N_2 are the count of total operators and operands respectively.

If the program is assumed to contain binary encoding then the size is defined as program volume:

$$\text{Program volume } V = N \times \log_2 n = \left(N_1 + N_2\right) \times \log_2\left(n_1 + n_2\right),$$

where $\log_2 n$ is used for the binary search method.

An algorithm can be implemented in various efficient and compact ways. The most competent and compact length of the program is defined as potential volume. For a program potential volume can be attained by specifying signature (name and parameters) of functions and subprograms previously defined and formulated as

$$\text{Potential volume } V^* = (2 + n_2{}^*) \times \log_2(2 + n_2{}^*),$$

where 2 represents the two operators (one for name of the function and other the separator used to distinguish the number of parameters) and $n_2{}^*$ represents the operand used for the count of input and output parameters.

Next Halstead defined the level of a program where level is the minimum possible size of the program. The level of a program having volume V and potential volume V* is defined as

$$\text{Program Level } (L)L = V^*/V,$$

where $0 \leq L \leq 1$, 0 denotes the maximum possible size and 1 denotes the minimum possible size of the program.

On the basis of the level of a program, Halstead defined the difficulty of writing a program as

$$D = 1/L,$$

where difficulty is the inverse of the program level.

Further he defined the effort metric to develop a program as

$$E = V/L = D \times V$$

As the volume and difficulty of the program increases, the development effort increases.

3.3.5 Key Findings:

- A range of complex metrics and their values are achieved using simple measures including operators, operands and size of the algorithm.
- There is no in-depth analysis requirement for the structure of the logic code; hence the ease of computation makes proposed metrics achievable and can be comfortably automated.

3.3.6 Metrics Used

- Operators and operands
- Functions and subprograms
- Input/output parameters

3.3.7 Factors Affecting Interaction and Integration Complexity

- Count of operators, operands, function names and similar measures.

3.3.8 Critique

- Originally software science was proposed to investigate the complexity of algorithms, not the programs, therefore these metrics are static measures.

- Halstead tested metrics on small-scale programs of even less than 50 statements. Applicability to large programs is questionable. These small-scale metrics cannot be generalized with respect to large, multi-module programs/software.

- In his theory Halstead calculated each occurrence of a GOTO statement as a distinct operator, whereas all the occurrences of an IF statement were treated as a single operator. Treating and counting different operators as different may create ambiguity.

Albrecht and Gaffney (1983) proposed the function-point technique to measure complexity in terms of size and functionalities provided by the system. This method addresses the functionality provided by the system from the user's point of view. To analyze the software system, Albrecht divided the system into five functional units on the basis of data consumed and produced. Three complexity weights, high, low and medium, are associated with these functional units using a set of pre-defined values. In function-point analysis (FPA), 14 complexity factors have been defined, which have a rating from 0 to 5. On the basis of these factors, Albrecht calculated the values of unadjusted function-point, complexity adjustment factors, and finally the value of function points (Pressman 2005).

Paradigm:

Conventional program/hierarchical software

Method:

FPA categorizes all the functionalities provided by the software in five specific functional units:

- *External inputs* are the number of distinct data inputs provided to the software or the control information inputs that modify the data in internal logical files. The same inputs provided with the same logic are not included in the count for every occurrence. All the repeated formats are treated as one count.

- *External outputs* are the number of distinct data or control outputs provided by the software. The same outputs achieved with the same logic are not included in the count for every occurrence. All the repeated formats are treated as one count.

- *External inquiries* are the number of inputs or outputs provided to or achieved from the system under consideration without making any change in the internal logical files. The same inputs/outputs with the same logic are not included in the count for every occurrence. All the repeated formats are treated as one count.

- *Internal logical files* present the amount of user data and content residing in the system or control information produced or used in the application.

- *External interface files* are the amount of communal data, contents, files or control information that is accessed, provided or shared among the various applications of the system.

These five functional units are categorized into three levels of complexity: low/simple, average/medium or high/complex. Albrecht identified and defined weights for these complexities with respect to all the five functional units. These functional units and corresponding weights are used to count the unadjusted function points:

$$\text{Unadjusted FP} = \sum_{i=1}^{5}\sum_{j=1}^{3}\left(\text{count of functional unit} * \text{weight of the unit}\right)$$

where i denotes the five functional units and j denotes the level of complexity.

Similarly, Albrecht defined the complexity adjustment factors on the basis of 14 complexity factors on a scale of 0–5. Adjustment factors provide an adjustment of ±35% ranging from 0.65 to 1.35. These complexity factors include reliable backup and recovery, communication requirement, distributed processing, critical performance, operational environment, online data entry, multiple screen inputs, master file updates, complex functional units, complex internal processing, reused code, conversions, distributed installations and ease of use. Complexity factors are rated as no influence (0), incidental (1), moderate (2), average (3), significant (4) and essential (5).

The complexity adjustment factor is defined as

$$\left[0.65 + 0.01 * \sum_{k=1}^{14}\left(\text{complexity factor}\right)_{i}\right]$$

The function point is now defined as the product of unadjusted FP and the complexity adjustment factor.

$$\text{FP} = \text{unadjusted FP} * \text{complexity adjustment factor}$$

3.3.9 Key Findings

- The function-point technique does not depend on tools, technologies or languages used to develop the program or software. Two dissimilar programs having different lines of code may provide the same number of function points.
- These estimations are not based on lines of code, hence estimations can be made early in the development phase, even after commencement of the requirements phase.

3.3.10 Metrics Used

- Count of inputs, outputs, internal logical files, external interfaces and enquiries.
- Weights of corresponding functional unit on the scale of low, medium and high.
- 14 complexity factors on the rating of values 0–5.

3.3.11 Factors Affecting Interaction and Integration Complexity

- Count of functions in the software.

3.3.12 Critique

- To compute the correct function-point count, proper analysis of requirements by trained analysts is required.
- Analysis, counts of functional units and computation of function points are not as simple as counting lines of code.

Henry and Kafura (1981) proposed a set of complexity computation methods for software modules/components. They suggested "software structure metrics based on information flow that measures complexity as a function of fan-in and fan-out." These metrics are based on the flow of data between the components of the application. Henry and Kafura defined the length of the module as the procedure length calculated using LOC or McCabe's complexity metric. This metric can be computed at a comparatively early stage of development. The authors used flows of local and global information from UNIX operating systems modules to define and validate their metrics.

Paradigm:

Conventional programs/hierarchical software

Method:

Henry and Kafura defined three categories of data flow in their work:

- **Global flow:** When a global data structure is involved between two modules. One module submits its data to the global data structure and the other accesses the submitted data from the data structure.
- **Direct local flow:** Flow of data between two modules is direct local if one module directly calls another module.
- **Indirect local flow:** Flow of data between two modules is indirect if one module uses data as an input returned by another module or both these modules were called by a third module.

Complexity metrics are defined on the basis of two types of information flow for a particular module or procedure

- **Fan-In:** Defines the sum of the number of local flows coming to the module and the count of data structures used to access the information.
- **Fan-Out:** Defines the sum of the number of local flows going from the module and the count of data structures modified by the module.

Henry and Kafura proposed the local flow complexity as "the procedure length multiplied by the square of fan-in multiplied by fan-out." This method is used to calculate the count of "local information flows" coming to (fan-in) and going from (fan-out) the module. That is:

Complexity in terms of local flows

$$= \text{length of the module}(\text{fan-in flows of the module} * \text{fan-out flows of the module})^2$$

High fan-in and fan-out values indicate high coupling among modules which leads to problems of maintainability.

Global flow complexity is defined in terms of possible read, write and read-write operations made by the procedures of the module. That is:

$$\text{Global information in terms of access and update} = (\text{write} * \text{read}) + (\text{write} * \text{read-write})$$
$$+ (\text{read-write} * \text{read}) + (\text{read-write} * (\text{read-write} - 1))$$

3.3.13 Key Findings

- The type, nature, number and format of the information which is going to transit among the software components are identified and defined well before actual implementation. These metrics can therefore be applied and estimated during the design phase.
- These design phase metrics can be used to identify shortcomings and flaws in procedure design construction and ultimately in modules.
- Through their metrics the authors argue that the size of the code plays a negligible role in complexity estimation.

3.3.14 Metrics Used

- Data and information transit among modules.
- Number of parameters used to access and to provide information.

3.3.15 Factors Affecting Interaction and Integration Complexity

- Number of incoming and outgoing flows.
- Number of parameters used to access and modify data structure.
- Number of operations updating the data structure.

3.3.16 Critique

- The length is computed using McCabe's formula or Halstead's formula, that is, the length of the code plays a vital role in the metrics.
- If the module has no interaction with other modules then the complexity of that module becomes zero.
- In the global information flow, only update operations participate in the complexity.

Chidamber and Kemerer (1994) proposed a metric suite for object-oriented software called the CK metrics suite. This metric suite is one of the most detailed and popular research works for object-oriented applications. The metric suite is defined in terms of complexity, coupling cohesion, depth of inheritance and response set, and is used to assess the complexity of an individual class as well as the complexity of the entire software system. In their metrics, Chidamber and Kemerer used the cyclomatic method for the complexity computation of individual classes.

Paradigm:

Object-oriented program/software

Method:

The authors define six object-oriented design metrics to analytically evaluate the complexity of software and programs. The metrics were tested on more than 2000 classes developed in C++ and Smalltalk. The authors consider the metrics available in object-oriented design inefficient, as they lack theoretical foundations. They list a set of six metrics:

- **Weighted methods per class:** Defined as the summation of the complexities of individual methods available in the class. That is

$$\text{Weighted methods per class} = \sum_{i=1}^{n} \left(\text{complexity of individual method}\right)_i$$

- **Depth of inheritance tree:** Defines the level of inheritance of methods from the deepest leaf node to the root in the class. It counts the length of the inheritance hierarchy.
- **Number of children:** Defines the count of descendent sub-classes directly belonging to a particular class.
- **Coupling between object classes:-** Denotes the number of objects that are coupled with a particular object.
- **Response for a class:** Defines the set of methods that belong to the class as well as the set of methods called by a particular method in that class.
- **Lack of cohesion in methods:-** Defines the level of cohesiveness among methods of a class.

3.3.17 Key Findings

- The count and the complexity of methods in the classes can be used to predict the time and effort required in pre- and post-implementation of the class.
- Availability of a greater number of methods in a class implies a smaller number of descendants of the class but it reduces the reusability feature of the class.
- As the depth of the inheritance tree increases, the scope of reusability will also increase. However, the longer length of inheritance hierarchy results in overheads for design complexity and testing efforts.

Abreu and Carapuca (1994) and Abreu and Melo (1996) proposed a metric set named "Metrics for Object-Oriented Design." In this metric suite, two fundamental properties of object-oriented programming are used: attributes and methods. Metrics proposed for the basic structural system of the object-oriented idea are encapsulation, inheritance, polymorphism and message passing. This suite consists of metrics for methods and attributes as an assessment method for encapsulation.

Cho et al. (2001) developed measures to quantify the quality and complexity of CBSE components. Using UML diagrams and source code, they define three categories of complexity measures: complexity, customizability and reusability of a component. Some of these measures are applicable to the design phase, while others can be implemented after the component installation phase. The argument is that the component should have customization properties in order to increase its reusability. The proposed metrics use McCabe's cyclomatic complexity and Albrecht's function points as the basis to compute the complexity and reusability of a particular program or method.

Paradigm:

Component-based software

Method:

Cho et al. categorized their quality estimation measures into three categories: complexity, customizability and reusability.

- **Complexity metrics:** Four classes of complexity metrics are proposed for components—plain, static, dynamic and cyclomatic.
- **Plain metrics:** These are defined on the basis of number of classes, abstract classes, interfaces, methods, complexities of individual classes, methods, corresponding weights, attributes and arguments.

Component plain complexity of a component

$= \big($number of external classes + summation of $\big($number of internal classes * weight of

corresponding classes$\big)\big)$ + number of in-out interfaces + $\big($number of external methods

+ summation of $\big($number of internal methods * weight of corresponding methods$\big)$

+ summation of complexity of classes in the component $\big($summation of count of

number of single attributes + summation of $\big($number of complex attributes * weight of

corresponding complex attributes$\big)\big)$ + summation of complexity of methods in each

class $\big($summation of count of number of single parameter + summation of $\big($number

of complex parameters * weight of corresponding parameter$\big)\big)$

These authors identify two types of classes: internal and external. Internal classes are defined in the component, whereas external classes are called from other components or libraries. Similarly, there are two types of methods: internal and external. Internal methods are defined within the class, whereas external methods are called from other classes. Weights are only assigned to internal classes and internal methods.

- **Static complexity metrics:** Static complexity is measured according to the internal structure of the component on the basis of associations among classes:

Component static complexity = summation of $\big($number of associations among classes * weight of corresponding association$\big)$.

Five types of association are identified and are assigned weights according to the order of their precedence in composition, generalization, aggregation and dependency. These associations are computed two classes at a time.

- **Dynamic complexity metrics:** Dynamic complexity is measured by taking the number of messages passed between the classes into account, within the component:

Component dynamic complexity

= summation of $\big($number of messages * frequency of messages$\big)$

+ $\big($summation of count of number of single parameter

+ complexity of each message$\big($summation of $\big($number of complex parameters *

weight of corresponding parameter$\big)\big)$

This metric is dynamic in nature since the number of parameters depends on the nature of execution.

- *Cyclomatic complexity metric*: Defined with the help of the source code developed. These authors used McCabe's cyclomatic complexity to assess the complexity of each method existing in a class.

Component cyclomatic complexity

$= \big($number of external classes $+$ summation of $\big($number of internal classes $*$ weight of corresponding classes$\big)\big) +$ number of in-out interfaces $+\big($number of external methods $+$ summation of $\big($number of internal methods $*$ weight of corresponding methods$\big)\big)$ $+$ summation of complexity of classes in the component $\big($summation of count of number of single attributes $+$ summation of $\big($number of complex attributes $*$ weight of corresponding complex attributes$\big)\big) +$ summation of complexity of methods in each class $\big($summation of count of number of single parameter $+$ summation of $\big($number of complex parameters $*$ weight of corresponding parameter$\big)\big) +$ summation of cyclomatic complexity of individual components $(e - n + 2)$

- **Customizability metrics:** Customizability is an attribute of a component that establishes its level of reuse. Three categories of customizable units in a method are identified and arranged in priority order as attribute, behavior and workflow. Corresponding weights are assigned to the behavior and workflow methods. To estimate the customization level, the suggested formula is

Component customization

$= \big($total number of methods including customizable attributes $+\big($number of methods including customizable behaviour $*$ weight of behaviour$\big)$ $+\big($number of methods including customizable workflows $*$ weight of workflows$\big)\big) /$ total number of methods available in the interface

- **Reusability metrics:** The reusability metric is defined at two levels: at component level, which assesses the reusability of a component in various applications; and at individual application level.

    Reusability at component level is estimated on the basis of methods providing common functionalities, as

Component reusability = total number of interface methods providing common functionalities / total number of interface methods available in the component

Reusability at the individual application level can be estimated using either lines of code or function points.

Component reusability (using lines of code) = number of reused lines of code of the components / total line of code in the application

Component reusability (using function points) = number of reused function points of the components / total line of function points available in the application

3.3.18 Key Findings

- The defined metrics cover both static and dynamic aspects of the component and the application, which are applicable to the design and post-implementation phases of the development.
- As the value of plain complexity increases, the value of a component's cyclomatic complexity increases. Dynamic complexity metrics exhibit more accurate results than static complexity metrics.
- Size, effort, cost and development time of components and component-based applications can be measured early and easily in the development phase.

3.3.19 Metrics Used

- McCabe's cyclomatic complexity.
- Alan Albrecht's function-point analysis.
- UML class diagrams, component diagrams and deployment diagrams.
- Structure of the code.

3.3.20 Factors Affecting Interaction and Integration Complexity

- Number of internal and external classes, internal and external methods, and in-out interfaces.
- Weights of internal classes, complex attributes, complex parameters and methods.
- Number of associations and their weights.
- Number and frequency of messages.

3.3.21 Critique

- Dynamic complexities are based on lines of code and function points. These metrics have their own problems and are heavily criticized by practitioners.
- It is not clear whether the weights associated with different entities during complexity estimations will be computed or assigned.

Narasimhan and Hendradjaya (2005) suggest a number of metrics to assess the complexity of component-based software. The packing density metric maps the count of integrated components, and the interaction density metric is used to analyze the interactions among components. Some constituents of the component are identified in their work, including line of code, operations, classes and modules. A set of criticality criteria for component integration and interaction are also suggested.

Vitharana et al. (2004), using what they term the "Business Strategy-based Component," developed a method for fabrication of components based on managerial factors like cost efficiency, ease of assembly, customization, reusability and maintainability. These are used to estimate such technical metrics as coupling cohesion, count, volume and complexity of components.

Lau and Wang (2007) argue that reusability is not only the purpose of component integration but is also a systematic software system construction process. To fulfil the basic objectives of CBSE, Kung and Zheng analyze an idealized component life cycle and suggest that components should be composed according to the life cycle. They also point out that the language of composition should have proper and compatible syntax.

Jain et al. (2008) assess the association and mappings of cause and effect among the system requirements, structural design and the complexity of the system integration process. They argue for fast integration of components so that the complexity impact of integration on architectural design of components can be controlled. Five major factors are used to analyze the integration complexity of a software system. These factors are divided into 18 sub-factors, including commonality in hardware and software subsystems, percentage of familiar technology, physical modularity, level of reliability, interface openness, orthogonality, testability and so on.

Parsons et al. (2008) propose specific dynamic methods for attaining and utilizing interactions among the components in component-based development. They also propose component-level interactions that achieve and record communications between components at runtime and at design time. The authors used Java components in this work.

Kharb and Singh (2008) propose a set of integration and interaction complexity metrics to analyze the complexity of component-based software. They argue that complexity of interactions have two implicit features: within the component, and from other components. Their complexity metrics include percentage of component interactions, interaction percentage metrics for component integration, actual interactions and interactions performed, and complete interactions in a component-based software.

Sharma and Kushwaha (2012) present an integrated method to assess development and testing efforts by analyzing the "improved requirement based complexity (IRBC)" in the context of component-based software.

A number of complexity assessment techniques for CBSE are suggested by academics on the basis of complexity properties including communication among components, pairing, structure and interface. The interaction and integration complexity measures available in the literature for development paradigms including conventional software and programs, object-oriented software, and component-based software, are summarized in Table 3.2.

The methods and metrics proposed so far in the literature are defined on the basis of interactions among instructions, operations, procedures and functions of individual and standalone programs and codes. These metrics are appropriate for small-sized codes. Some measures are also defined for object-oriented software, but they are not adequate for CBSE applications. In CBSE, components exchange services and functionalities with each other through connections and communications. Interaction edges denote the connections among components, with an edge for each requesting communication and an edge for

TABLE 3.2

Summary of Interaction and Integration Complexities

Paradigm	Measures and Metrics Used	Key Findings	Factors Affecting Interaction and Integration Complexity	Author(s)/ References
Conventional software and programs	• Line of code • Interaction among statements • Nodes and interactions	• Control-flow graph of a program used to compute the cyclomatic complexity • Graph-theoretical notation used to draw control-flow graph where a graph G has n vertices, e edges and p connected components	• Conditional statements • Loop statements • Switch cases	McCabe (1976)
Convention-al software and programs	• Line of code • Operator count • Dissimilar operands count • Total count of dissimilar operators • Total count of dissimilar operands	• Proposed a complete set of metrics to measure the complexity of a program considering various factors, like program vocabulary, program length, program volume, potential volume and others	• Program vocabulary • Program length • Program volume • Effort • Time	Halstead (1977)
Modular programming	• External inputs • External outputs • External enquiries • Internal logical files • External interface files	• Proposed function-point analysis technique to measure size of a system in terms of functionalities provided	• 5 functional units • 14 Complexity factors • Complexity adjustment factors • Degree of influence	Albrecht and Gaffney (1983)
Modular programming	• Fan-in information • Fan-out information • Complexity of the module • Line of code • McCabe cyclomatic complexity	• "Software structure metrics based on information flow that measures complexity as a function of fan-in and fan-out."	• Number of calls to the module • Number of calls from the module • Length of the module	Henry and Kafura (1981)
Object-oriented software	• Methods • Inheritance • Coupling • Cohesion • Object library effectiveness • Factoring effectiveness • Method complexity • Application granularity	Object-oriented metrics to assess complexity and productivity metrics, including average number of methods per object class, inheritance tree depth, average number of uses dependencies per object, arcs and degree of cohesion of objects	• Maintainability • Reusability • Extensibility • Testability • Comprehensibility • Reliability • Authorability	Morris (1989)

Type		Description		Reference
Object-oriented software	• Lines of code to count the size • Number of screenshots • Number of reports	• Object-point metric computed using counts of the number of screenshots, reports and components based on their complexity levels • Complexity levels classified as simple, medium or difficult	• Line of code • Complexity levels	Boehm (1996)
Object-oriented software	• Cyclomatic method • Class complexity • Methods • Object-oriented properties	• One of the most detailed and popular research works in object-oriented software, including weighted method per class, depth of inheritance tree, number of children, lack of cohesion in methods	• Complexity • Coupling • Cohesion • Inheritance • Number of children • Response set	Chidamber and Kemerer (1994)
Object-oriented software	• Method-hiding factor (MHF) • Attribute-hiding factor (AHF) • Method-inheritance factor, • Attribute-inheritance factor for inheritance • Polymorphism factors • Coupling factors	• Uses two fundamental properties of object-oriented programming; attributes and methods • MHF and AHF together are measure of encapsulation	• Encapsulation • Inheritance • Polymorphism • Message passing	Abreu and Carapuca (1994)
Component-based software	• Levels of complexity • Quality of components • Customizability	• Metrics to measure quality and complexity of components • Mathematical equations and expressions used in metrics	• Size • Costs • Effort • Reuse level	Cho et al. (2001)
Component-based software	• Indicates these values as high or low • Establishes a relationship among these proposed metrics	• Hierarchical model consisting of three layers: quality, criteria and metrics	• Understandability • Adaptability • Portability	Washizaki et al. (2003)
Component-based software	• Line of code Operations • Classes • Modules • Number of components	• Two sets of metrics to assess the complexity of component-based software • Two complexity metric suites: component packing density metrics; component interaction density	• Risk associated with components • Constituents • Interactions among components	Narasimhan and Hendradjaya (2005)
Component-based software	• Coupling • Cohesion • Number of components • Component size • Complexity	• Methodology for fabrication of components	• Syntax • Semantics	Vitharana et al. (2004)

(continued)

TABLE 3.2 (*Continued*)
Summary of Interaction and Integration Complexities

Paradigm	Measures and Metrics Used	Key Findings	Factors Affecting Interaction and Integration Complexity	Author(s)/ References
Component-based software	• Prioritization of requirements • Functional modularity • Feasibility • Interface • Testability	• Assesses association and cause and effect mappings between system requirements, system architecture and systems integration complexity procedure • Five major factors to analyze integration complexity of software system • These factors are divided into 18 sub-factors	• Commonality in hardware and software subsystems • Percentage of familiar technology • Physical modularity • Level of reliability • Interface openness • Orthogonality, testability	Jain et al. (2008)
Component-based software	• Static interaction complexity • Dynamic interaction complexity	• Specific dynamic methods for attaining and utilizing interactions among the components in component-based development • Component-level interactions that achieve and record communications between components at runtime and design time	• Call traces • Call graphs • Runtime paths • Calling context trees	Trevor et al. (2008)
Component-based software	• Interface • Implementation • Deployment • Incoming and outgoing interactions	• Set of integration and interaction complexity metrics to analyze the complexity of component-based software, including percentage of component interactions, interaction percentage metrics for component integration and actual interactions	• Maintainability • Reusability • Reliability	Kharb and Singh (2008)

each responding communication. However, practitioners and researchers do not include both edges in their complexity computations. Rather, single-edge theory, which does not apply to CBSE, is used in graph representations and all assessments (Gill and Balkishan 2008, Tiwari and Kumar 2016).

3.4 Complexity Issues

The Cambridge Dictionary defines the term complexity as the "state of being formed of multiple parts or things and that are difficult to understand." In the context of algorithms and programs, it is defined thus: "estimation and prediction of resources acquired by a solution (algorithm or program) to complexity is identified as an indirect measurement." (Tiwari and Kumar 2014). It cannot be calculated as a direct measurement like lines of code or cost. Complexity is the property of a system that makes it difficult to formulate its overall behavior in a given development environment. Software complexity is a term that encompasses numerous properties of a piece of software, all of which affect internal as well as external interactions. Complexity may be characterized as computational, algorithmic or information processing. Computational complexity focuses on the amount of resource required for the execution of algorithms like space and time, whereas information-processing complexity is a measure of the total number of units of information transmitted by an object.

In component-based software engineering, complexity is a measure of the interactions of the various elements or components of the system. Complexity describes interactions among entities like functions, modules and components. As the number of entities increases, the number of interactions between them increases according to the software requirements.

Common major issues regarding complexity of component-based software are:

- **Terminology for defining complexity and related terms:** Specific and appropriate vocabulary is required to define terminology related to complexity in the context of component-based software rather than applying meanings of complexity taken from other fields.

- **Identifying and defining complexity factors of components and component-based software:** Some factors increase and some decrease the complexity of software. Proper mechanisms should be used to identify and address factors that affect the complexity of components as well as the complexity of component-based software.

- **Techniques are required to assess complexity of individual components:** Components produce complexity due to intra-module interactions which may be necessary to make the component usable. The use of complex components will increase the complexity of the overall system. There should be an appropriate method to assess the complexity of the component. The concept of complexity applies equally to both types of component, black-box and white-box.

- **Metrics are required to evaluate the complexity of component-based software:** Complexity measures for component-based software are as important as the measures for individual components. Component-based development focuses on inter-component interactions. These inter-component interactions generate new complexity aspects that do not exist in traditional or small-scale software or applications. New measures and metrics are required to address this new problem domain.

3.5 Design and Coding Issues

Design is the most creative phase of the development of software constructs and products. Large-scale software like component-based software needs proper architectural design before actual development commences. Designing multi-component software is a complex and chaotic task. Knowledge of the requirements provided by the customer or documents made available by system analysts is not sufficient. The availability of expertise in designing tools and methodologies is an additional benefit but does not address core design issues.

Similarly, implementation of designed software in terms of code is a demanding activity that requires time and patience. There is a common saying that "design is not coding and coding is not design." Good design does not imply better code implementation. Coding of components and component-based software needs a different approach both conceptually and practically. Expertise in programming languages or proficiency in development tools is not adequate for coding such complex and multifaceted software design. Complications may increase dramatically if faults from previous phases are identified during the coding phase.

In view of these design and coding concerns, the following issues must be addressed in the context of component-based software development:

- **Identification of core requirements of the software:** When customers provide the list of requirements, analysts analyze them for feasibility-related issues. Designers must be aware of the core requirements of the software which should be resolved at initial meetings. These core requirements are either identified in customer and analyst meetings or should be outlined by the designer.

- **Levels of design:** Designing is a complex and multi-part activity which generally requires time and proper scheduling of other activities. Designing can be divided into two levels: system design and detailed design. System design consists of outlining the solution, identifying major modules and components according to the functionalities required, formation of teams, identification of tools and resources, and similar activities. Detailed design includes implementation details such as the working of each module and component, integration of different modules, proper algorithms, database design and similar tasks.

- **Concept of componentization:** Component-based software is divided into components to achieve the desired functionalities. Once components are identified and assembled, the level of componentization must be determined so that components can be managed. If designers try to keep the number of components down, it may result in increased component size, but reducing the size of components may increase their number. The trade-off between size and number of components should be properly addressed.

- **Component design issues:** Components may be developed from scratch according to the requirements of the software. Designing such components is a challenge, as they may be integrated with existing components. Designing context-free software which can be deployed anywhere with any component is challenging.

- **Parallel development of components:** Component-based software development focuses on parallel development of components. This development requires the proper design and scheduling of software architecture. Parallel development reduces overall cost, development time and resource required.

- **Component-based software design issues:** Designing large and complex component-based software is a multifaceted activity which requires knowledge of the problem domain, expertise in design tools and skills, and experience of designing large-scale software. Component-based software requires unambiguous architecture as well as design of each and every detail of the software constructs.

- **Component coding issues:** Coding of individual components must be performed according to the software requirements. Components are developed with many factors in mind, including reusability, quality, reliability, understandability, ease of use, composability and adaptability. Components are developed in such a way that they can be independently deployed in varied environments to perform the desired functions.

- **Component-based software coding issues:** Implementing multi-component software into code is a very challenging job requiring advanced knowledge and coding skills. Some new components are developed and integrated with already available components.

3.6 Testing Isssues

Testing plays a vital role in the development of components and component-based software engineering. Crucial issues in testing components and component-based software are:

- **Testing techniques/metrics for black-box components:** There may be components whose source code may not be available; we call such components black-box components. Techniques for testing black-box component behavior are required after providing inputs or after integration with other components in component-based software.

- **Testing techniques/metrics for white-box components:** White-box components are those whose source code is available. Efficient testing measures of white-box components' behavior are required after providing inputs or after integration with other components in component-based software.

- **Measures and metrics to assess testing effort:** Measuring the effort required for testing is an important issue in the component-based software environment, as testing is one of the most crucial phases. Reusability of the component increases as the testing effort decreases.

- **Measures and metrics to assess testing cost:** The overall software cost depends on various factors. The testing cost is one of the vital factors. It is important for developers as well as for customers to assess the cost of testing in advance. It enhances the choice of reusability of the component.

- **Measures and metrics to assess testing duration:** Testing is one of the most time-consuming activities in the development of a component and component-based software. Measures and metrics to evaluate testing time of components as well as of the complete software are heavily in demand.

- **Test-case generation techniques for individual components:** Test-case generation methodologies for individual components are one of the least focused areas in the component-based development research field. There are many testing techniques for components but efficient test-case generation techniques are not available. There is great scope in this domain.

- **Test-case generation techniques for component-based software:** For conventional software there is a set of testing techniques and methods, including black-box, white-box and other techniques. But for component-based software there are no clearly specified test-case generation techniques.

- **Test-case minimization techniques/metrics during integration testing in component-based software:** As regression testing takes place during integration of components, a great many test cases are produced. During integration testing, the number of test cases increases incrementally because of repeated test cases. As the components are assembled, new test cases in addition to repeated test cases increases the overall number of test cases. Techniques to minimize the number of test cases are needed.

- **Test-case minimization techniques/metrics during system testing in component-based software:** After component assembly according to the specified architecture and integration testing, system testing commences, which requires new test cases. The focus should be on minimizing the number of test cases in this phase.

- **Test-case minimization techniques/metrics during other testing in component-based software:** Various components require various testing techniques in addition to system testing. Every testing process requires and produces new and sometimes repeated test cases. There is a need for measures that emphasize minimization of test cases in these testing activities.

Testing is one of the crucial phases of the overall development of software. Testing verifies the correctness, precision and compatibility of the software at the individual as well as the system level. Practitioners have identified that improper testing results in unreliable products (Elberzhager et al. 2012, Tsai et al. 2003). In today's software development environment, testing commences just after the finalization of system requirements.

In component-based software engineering, testing starts from the component level and moves on to the integrated CBS system level (Pressman 2005). Testing is commonly used to *verify* and *validate* software (Gao et al. 2003, Myers 2004, Weyuker 1998). *Verifying* components in CBSE is the collection of procedures that certify the functionalities of components at individual level. *Validating* components is a series of procedures to ensure the integrity of integrated components according to the architectural design while fulfilling the needs of the customer. In the literature, there are two groups of testing techniques, black-box and white-box.

3.6.1 Black-Box Testing

Black-box testing methods focus on assessing the functional behavior of the software through inputs provided and outputs observed. Black-box testing treats the internal logic of code as a black box and the testing observations are captured using inputs and outputs only.

Black-box testing strategies proposed in the literature (Ntafos 1988, Ostrand and Balcer 1988, Ramamoorthy et al. 1976, Voas 1992 Voas and Miller 1992, 1995, Weyuker 1993) are:

  i. Boundary-value analysis (BVA)
 ii. Equivalence-class partitioning (ECP)
iii. Decision table-based testing (DTB)
 iv. Cause-effect graphing (CEG)

3.6.1.1 Boundary-Value Analysis (BVA)

In boundary-value analysis (BVA) testing, the emphasis is on verifying that the software behaves normally (accurately) at the extreme boundaries (Ramamoorthy et al. 1976, Voas 1992), and the assumption is that a large number of errors occur at the boundaries rather than within the limits of the conditions. The boundary-value analysis method is used to draw test cases focusing on the input values at the edges of the logical conditions (Voas and Miller 1992). This method usually tests the center values, boundary values, and the values just below and just above the boundary.

3.6.1.2 Equivalence-Class Partitioning (ECP)

Equivalence-class partitioning (ECP) is a black-box testing strategy based on the assumption that if we divide the input values into classes or partitions on the basis of their validity or invalidity, then the impact of each value under the particular class will be equivalent. That is, the validity of every data member of the legitimate class is equal or similar; and every data member of an illegitimate class is equally invalid. In ECP the input domain is partitioned and tested for validity. ECP is usually applied to input classes, but sometimes also to output classes (Ntafos 1988, Ostrand and Balcer 1988).

3.6.1.3 Decision Table-Based Testing (DTB)

The decision table-based (DTB) testing method is used to test the permutations of input conditions rather than testing single input values. Decision table-based testing was proposed to overcome the limitations of BVA and ECP, which are only applicable when single values of input and output domain are to be tested. The DTB testing method combines two or more input logical conditions, and assesses and examines these complex conditions. There are four quadrants in the decision table: condition stub combines the input conditions; action stub specifies the output conditions; condition entries inputs condition entries; and action entries inputs action entries (Table 3.3).

TABLE 3.3

Decision Table

The four quadrants	
Condition stub	Condition entries
Action stub	Action entries

In the DTB method, the condition stub and condition entries correspond to an input condition or set of input conditions. The action stub and action entries correspond to an output or set of outputs (Voas and Miller 1995, Weyuker 1993).

3.6.1.4 Cause-Effect Graphing

The cause-effect graphing method is also used to combine the input conditions so that more elaborate assessment can be performed. In this technique, inputs are recognized as causes and outputs as effects. A Boolean graph known as a cause-effect graph is drawn by joining these causes (inputs) and effects (effects). It is a simple graphing technique where nodes represent input conditions and edges represent linking between nodes.

3.6.2 White-Box Testing

White-box techniques are used to address the testing requirements of the structural design and internal code of the software. White-box testing methods ensure that all execution paths and independent logics of the program or module have been tested adequately. The logical decisions given in the code must also be tested on their true and false sides, and all looping and branching codes at the extremes as well as within the boundaries must be checked at least once. *Basis-path testing* is used in the white-box technique to analyze the structure of the software.

McCabe (1976) proposed a method to count the number of test cases of a program. He proposed cyclomatic complexity based on the structural design of the code. He defined a control-flow graph having n nodes, e edges and p connected components. The cyclomatic complexity is calculated as $V(G) = e - n + 2p$; here 2 is the "result of adding an extra edge from the exit node to the entry node of each component module graph" (Chen 2011).

Henderson-Sellers (Henderson-Sellers and Tegarden 1993) proposed an amendment to McCabe's formula to compute cyclomatic complexity. Henderson-Sellers' definition, "$V(G) = e - n + p + 1$," used the concept of modularity. The altered formula argued that to make components strongly linked, an edge is added, denoted as constant 1 to the multi-component flow graph. Orso et al. (2001) suggested a technique using component metadata for regression testing of components and their interfaces. This metadata consists of data and control dependencies, source code, complexity metrics, security attributes, information retrieval mechanisms and execution procedures. They also proposed a specification-based regression test selection technique for CBS.

Michael et al. (2002) proposed that the system's effectiveness testing can be increased by regulating certain component factors. The factors eligible for optimization are cost, reliability, effort and similar attributes. Single as well as multiple application systems are considered for "software component testing resource allocation," and "reliability-growth curves" are used to model the association between failure rates and "cost to decrease this rate." Interaction among components and failure rates of components are used in this methodology.

Bixin et al. (2005) developed a matrix-based method to compute the dependencies in component-based software. Categories of dependencies available in component-based software are identified, and a "component-dependence graph" and "dependence matrix" are used to record them. On the basis of the dependence graph and dependence matrix, a mathematical foundation is proposed to assess these dependencies in component-based software.

Dallal and Sorenson (2006) introduced a method called "all-paths-state" to produce state-based test cases through testing at class level. They identified framework interface classes to cover the maximum number of specifications. It is suggested that this technique will help to generate reusable test cases effectively, and that the framework design is helpful for both reusable and newly developed modules.

Gill and Tomar (2007) presented work focusing on the testing requirements and test-case documentation process for component-based software. The testing requirements proposed are access to the source code, functions provided by the component, compatibility with other components, component middleware, interactions of various reusable components and component specifications. Testing limitations for component-based software are also discussed.

Sen and Mall (2012) proposed a testing technique based on the state and dependence of components. The technique is called a regression test selection technique and targets the regression test suite. It is suggested that this method will reduce the number of test cases in regression testing, and the authors assume a very strong link between the state model in the design phase and the executable code, whereby the state model design is maintained and the code is altered without affecting the state of the component.

Fraser and Arcuri (2013) proposed an approach covering all testing goals at the same time, leading to a more effective result in terms of minimizing test suites. Object-oriented software was used in this work.

Edwards (2000) describes test-case methods for black-box and white-box components in CBSE. This technique relied on the concept of generating a flow graph from the component specification, and then applying conventional graph coverage methods. Pre-condition, post-condition and fault-detection ratios were used in the test-case design.

Hervé et al. (2013) propose a technique for building and installing reusable integration connectors to solve problems arising during the integration of off-the-shelf components. These problems are defined in the form of exceptions that can be healed automatically by the connectors. The connectors are usually installed on the basis of integration information available with the components. The complete structure, exception behavior and integration behavior of these healing connectors is provided.

Table 3.4 summarizes testing issues in component-based software.

In component-based development, software systems are partitioned into a number of small and manageable components, which changes the architectural design as well as the behavioral attributes of the software. To adapt to these changes, various categories of components are engineered and used in varied frameworks. Afterwards, currently engineered components and existing components are assembled according to a pre-planned architecture to contribute their functionalities. Very few of the techniques defined in the literature are applicable to scenarios with two or more components. To calculate the number of test cases in white-box testing techniques like cyclomatic complexity are used (though only in the context of individual components). In black-box testing, techniques that can calculate the number of test cases for component-based software are required (Tiwari and Kumar 2014).

TABLE 3.4

Summary of Testing Issues in Component-Based Software

Paradigm	Measures and Metrics Used	Key Findings	Testing Factors	Author(s)/ References
White-box testing	• Statements of code • Interaction among statements • Control-flow graph	• Graph-theoretic notations to draw the control-flow graph and formula for cyclomatic complexity • Formula gives the count of test cases and computes number of independent logics in the program code • Vertices denote the instruction of the code and edges present flow of controls among vertices	• Conditional statements • Loop statements • Switch cases	McCabe (1976)
White-box Testing	• Statements of code • Interaction among statements • Multi-component flow graph	• Modified formula for computing cyclomatic complexity with the argument of an extra edge	• Conditional statements • Loop statements • Switch cases	Henderson-Sellers and Tegarden (1993)
White-box testing	• Source code • Interaction among components • Components specification	• Regression testing technique using component metadata • Regression test method using software specification	• Metadata including data and control dependencies • Source code, complexity metrics • Security attributes	Orso et al. (2001)
Black-box testing	• Inter-component interaction • Single application • Multi-application systems	• Methodology to optimize testing schedules, subject to reliability constraints • Generates optimization opportunities in testing phase • Effectiveness of the system testing can be increased by regulating component factors	• Cost • Reliability • Failure rate • Testing schedule	Michael et al. (2002)
White-box testing	• Dependency matrix • Interaction • Source code • Adjacency matrix	• Component-matrix graph and dependence matrix • proposed to assess dependences in component-based software Categories of dependencies defined including state, cause and effect, input/output context and interface-event dependence	• Data • Control • Time • Context • State	Bixin et al. (2005)
Black-box testing	• All-transition coverage • Pair coverage • Full predicate coverage • Round-trip path coverage	• Testing method of all-path state to reduce number of test cases • Black-box and white-box testing methods	Reusable and newly developed classes	Dallal and Sorenson (2006)

White-box testing	• Source code • Interactions • Specification of components	• Focused on two aspects of testing component-based software: testing requirements and test-case documentation process • Source code, functions, compatibility, middleware, interactions and specifications as testing requirements • Testing process defined as test strategy, planning, specification, execution, recording, completion and test results	Guidelines and testing process step-wise and phase-wise	Gill and Tomar (2007)
White-box testing	• Dependency graph • Source code • Component state • Reverse engineering	• Regression selection technique to minimize count of test cases in component-based software • Algorithm to automatically generate code from state model	• Method parameters • Operations • Conditions (like looping and branching)	Sen and Mall (2012)
Black-box testing	• EVOSUITE tool • Line of code • Number of test cases	• Whole coverage testing technique to cover all testing goals at same time rather than one at a time • Approach claimed to minimize number of tests • Study uses primitive, constructor, field method and assignment statement	• Object-oriented classes • Member variables • Numeric, Boolean, string • Enumeration variables • Test suites	Fraser and Arcuri (2013)
White-box testing	• Pre-condition • Post-condition • Fault injection • Mutants • Test cases	• Specification-based test-case generation technique • Flow graphs defined with the help of specifications	• Component specification • Enumerating paths	Edwards (2000)
White-box testing	• Integration among components • UML modeling • Reusability	• Concept of healing connectors from developer point of view • Problems defined as exceptions • Connectors usually installed on the basis of integration information available with components	Proved through case studies	Chang et al. (2013)

3.7 Reliability Issues

Major issues for reliability of component-based software are:

- **Reliability assessment metrics:** A large proportion of reliability estimation methods defined in the literature are either hardware oriented or conventional software/program oriented. As component-based software components share interactions and integrations, these complexities, together with reusability issues, have to be considered when assessing the reliability of CBS.

- **Factors affecting reliability:** Reliability of component-based software depends on various factors, the identification and definition of which is a prominent issue in component-based software research.
- **Role of reusability in reliability:** While many eminent researchers have proposed reliability estimation metrics, the basic feature of component-based software—reusability—is rarely considered. Reusability is in fact a major reliability factor.

Reliability is key to a system's quality. Reliability defines not only the correctness but also the precision attributes of software. The literature includes a number of quality reliability models, and reliability assessment methodologies have been invented to predict the reliability of CBSE applications. In the development of quality software, various independent components have to be integrated according to the architectural design of the software. It is not sufficient to apply traditional reliability metrics when measuring the reliability of such software and in a heterogeneous environment. Different researchers have categorized software reliability in diverse ways. Eusgeld et al. (2008) identify three types of software reliability:

a. Black-box reliability
b. Metric-based reliability
c. Architecture-based reliability or white-box reliability

    a. *Black-box reliability* (Farr 1994, Goel 1985, Ramamoorthy and Bastani 1982) assesses reliability using failure observations of the software over a period of time during testing. These reliability estimation models treat the system as a whole rather than treating its internal structure and intra-component interactions.

    b. *Software metric-based reliability* (Musa 1998) estimates the reliability of the software by analyzing the its static properties, such as lines of code, asymptotic complexity, developer experience, software engineering methods and techniques used for testing.

    c. *Architecture-based reliability* (Krishnamurthy and Mathur 1997) estimates reliability by considering the internal structure of the software. During reliability prediction and assessment, integration and interaction among components is represented by architecture. While calculating the reliability of the software, component reliabilities are also considered.

Goseva-Popstojanova and Trivedi (2001) categorized all architecture-based reliability estimation methods into three broad categories:

a. State-based estimation methods
b. Path-based models
c. Additive methods

    a. *State-based estimation methods* (Gokhale et al. 1998, Kubat 1989, Littlewood and Strigini 1993) use a control graph to show the software architecture as well as the interaction among components. In state-based models, the basic assumption is that components fail independently. Components' states follow Markov behavior, i.e., the current state and the past states of a component are independent of each other.

    b. *Path-based models* (Goseva-Popstojanova and Trivedi 2001, Krishnamurthy and Mathur 1997, Shooman 1997) estimate reliability by considering the possible execution paths of the program. From starting component to ending component, every possible path is extracted.

c. *Additive methods* (Everett 1999, Xie and Wohlin 1995) use a non-homogeneous poisson process (NHPP) to model the reliabilities of the components. In these models, software failure is represented in NHPP terms through the number of component failures and the intensity functions of failures of individual components.

Littlewood's architecture-based model (Littlewood 1975) is one of the early models developed for reliability estimation of intra-modules and inter-modules. This architectural model suggests that components as well as interfaces can fail with constant failure rates. In this model, a continuous-time Markov chain is used to estimate software reliability.

Cheung's reliability model (Cheung 1980), a state-based model, uses a control-flow graph to represent the component's architecture. A transition probability matrix is drawn to show control transfer among components.

Kubat's model (Kubat 1989) uses execution times of components. In this estimation method, a discrete-time Markov chain (DTMC) represents the transition between components. The probabilities of execution of components are taken into account during reliability estimation of CBSE applications.

Gokhale (Gokhale et al. 1998) used random production of faults in components to estimate the failure of components at arrival. In this technique, reliability is estimated through simulation by counting the number of failures of components. Here the basic assumption is the availability of failure rates and repair rates of components.

Sanyal (Sanyal et al. 2004) proposed a reliability estimation technique based on a component dependency graph and fault propagation delay. A reverse engineering technique was used along with the source code, assuming that the logic and code of the component is accessible. This approach may not be helpful for black-box components or components whose code is not accessible.

Krishnamurthy and Mathur (1997) developed a path-based reliability model based on test data, test cases and execution paths of the application's architecture, using the theory that components fail independently and the interfaces are error free. This method computes path reliabilities by executing the sequence of components one after the other and then averaging all of them to assess the reliability of the software.

Everett's reliability estimation model (Everett 1999) assesses component reliability using an extended execution growth model. The processing time of each component is tracked to estimate the reliability of CBSE applications. This method infers the parameters from the component's attributes and from the knowledge of test cases as well as operational usage. Thus, this technique considers the execution time of individual components. Here the total failure count as well as failure intensity is added as a function of reliability of the software.

Singh et al.'s reliability prediction model (Singh et al. 2001) uses unified modeling language to analyze the reliability of component-based software in work which also integrates UML models and failure rates of components. Scenarios and case studies are used to estimate reliability in the initial software construction phases. An algorithm is proposed to predict the reliability of CBSE applications, considering the number of components and number of scenarios and their probability of execution.

Cortellessa et al. (2002) suggest a reliability assessment technique for component-based software. The goal is to collect component failure rates and connector failure rates. The assumption is that every component's individual reliability is available. UML diagrams, deployment diagrams and sequence diagrams are used to denote communication between components. This technique also assumes that components fail without affecting each other.

Reussner et al. (2003) propose a reliability assessment technique that analyzes usage profiles and the reliability of the component's environment. Component reliability can be calculated as a factor of the usage profile, including the reliability of external services. This technique is applicable to black-box components. This reliability prediction method is applied to open and distributed systems. In this method components follow the Markov chain property.

Yacoub et al. (2004) propose a "scenario-based" reliability estimation approach to the reliability of CBS. Dependency graphs are used and algorithms proposed that compute the reliability of CBSE applications. Algorithms are extended to cover distributed components and the hierarchy of subsystems for reliability estimations.

Rodrigues et al. (2005) offer a reliability estimation method that considers the structure of the component and the scenarios present in the component-based application. This method does not calculate a component's reliability, but assumes that the component's reliability is given. In this technique, components follow Markov properties. It is assumed that component failures are independent of each other.

Gokhale and Trivedi (2006) propose reliability predictions for component-based software taking its design structure and the failure behavior of the components into account. A unifying framework to predict reliability of state-based models as well as architecture-based software is proposed. Software architecture is considered alongside component failure behavior and interfaces to produce a composite method which is analyzed to predict the reliability of the application.

Sharma and Trivedi (2007) developed an "architecture-based unified hierarchical model" to assess and predict the performance of the software, the reliability of the application, security and cache behavior of the components. DTMCs are used to model software and equations are defined that assess the software application on its architectural design and the behavioral properties of the standalone components. Sharma and Trivedi's hierarchical model covers reliability, performance, security and "cache-miss" behavior predictions.

Zhang et al. (2008) suggest an approach based on a component-dependence graph. This is a sub-domain technique based on an architectural-path reliability model; algorithms are suggested to estimate the reliability of the CBSE applications. The software operational profile is assumed to be known. Control is also assumed to flow from one component to another. This method is capable of categorizing software application reliability on the occasion of a change in the operational profile.

Hsu et al. (Hsu and Huang 2011) developed an adaptive method using path-based prediction for complex CBS systems. They used three techniques to estimate path reliability of the whole application: sequence, branch and design of loops. This method also suggests the effect of failure for every component in the overall reliability of the application.

Palviainen et al. (2011) proposed an assessment method for the prediction of reliability of components and CBSE applications. A method for reliability prediction of components under development is defined, initially during the software development phases. This method also suggests the effects on reliability of selection of the correct components. Heuristic, model-driven reliability prediction is combined with component-level and system-level techniques to explore the development of reliable CBS applications.

Fiondella et al. (2013) suggest a reliability assessment technique that uses "correlated failures." Component reliability is estimated by considering reliabilities of individual components, correlation of failures and the architecture of the software. This method follows a multivariate Bernoulli distribution to compute overall component-based software reliability.

Reliability methods available in the literature are explored by considering the path-based estimation paradigm and summarized in Table 3.5.

TABLE 3.5

Summary of Reliability Issues in Component-Based Software

Paradigm	Measures and Metrics Used	Key Findings	Reliability Estimation Factors	Author(s)/ References
Path-based	• Dependence graph • Components	• Proposes program dependency graphs and fault propagation analysis for analytical reliability estimation of component-based applications • Code-based approach (reverse-engineering) where dependency graphs are generated from source code • Not applicable for off-the-shelf components	• Source code • Interactions	Sanyal et al. (2004)
Path-based	• Test cases • Components • Intra-component dependencies	• Proposed method estimates path reliability based on the sequence of components executed for each test run and averages them over all test runs to obtain an estimate of system reliability • Method is based on test data, test cases and execution paths of the application's architecture • Assumption based on theory that components fail independently and interfaces are error free	• Components • Interfaces • Execution paths • Component reliability • Execution sequence of components	Krishnamurthy and Mathur (1997)
Scenario-based (Path-based)	• UML diagrams • Use-case diagrams • Sequence diagrams • Independence failure • Regularity failure • Number of components	• Proposes a reliability estimation technique using unified modeling language • Pre-defined scenarios and case studies predict reliability in early phases of development • Proposed algorithm considers number of components and number of scenarios and the probability of their execution	• Reliabilities of components • Operational failures • Probability of execution of use case • Number of sequence diagrams	Singh et al. (2001)
Path-based	• Use case diagrams • Sequence diagrams • Deployment diagrams • Calculating formulas	• Estimation method for reliability of component-based software collects component and connector failure rates • Method assumes that individual component's reliability is known • For component modeling UML diagrams are used • Method estimates that various components fail independently	• User profiles • Component's reliability • System failure properties	Cortellessa et al. (2002)

(continued)

TABLE 3.5 (*Continued*)

Summary of Reliability Issues in Component-Based Software

Paradigm	Measures and Metrics Used	Key Findings	Reliability Estimation Factors	Author(s)/ References
Path-based	• Control instructions • Kens • Markov property • Probability of call of components • Service reliability • Overall reliability	• This method estimates reliability of component-based software by composing profiles of component usage and reliability of environment • Method is applicable to black-box components whose code is not available • Components fail independently	• Usage profiles • Reliability of components • Failure rates	Reussner et al. (2003)
Scenario-based (Path-based)	• Component dependency graph • Interaction among components • Transition probability • Transition reliability	• Scenario-based reliability estimation method for component-based software • Assumes that profiles of execution scenarios are available • Stack-based algorithm assesses reliability • Methodology can be applied in early phases of software development since execution scenarios are designed in the design phase	• Individual reliabilities of omponents • Execution times of components • Link reliabilities • Execution scenarios • Execution time of a scenario	Yacoub et al. (2004)
Path-based	• Execution scenarios • Multiple scenarios, • Markov property • Scenario specification • Interaction among components	• Given a reliability estimation method considering the structure of the component and the scenarios present in component-based application • Method is not for component reliability, but assumes that the component's reliability is given • Interactions among components are taken in general • Method follows Markov property • Assumes that component failures are independent	• Operational profiles • Scenarios • Component reliability	Rodrigues et al. (2005)
Path-based	• Individual component reliabilities • Failure rates of individual components, • Number of visits of a component	• Reliability prediction method named "Composite Method" combines application architecture with failure behavior of components and interfaces into a composite model that can be analyzed to predict the reliability of the application	• Probability of transition from one component to another • Reliability of components • Expected number of visits • Expected time	Gokhale and Trivedi (2006)

Path-based	• Performance prediction • Identification of performance bottlenecks • Performance attributes of each component • Reliability prediction • Identification of reliability bottlenecks • Security prediction • Identification of security bottlenecks	• Architecture-based unified hierarchical model to assess and predict software performance, application reliability, security and cache behavior of components • Method of security prediction and cache-miss analysis takes software architecture into consideration • This model is fairly accurate as it takes into account the second-order architectural effects	• Variation in the number of visits to each component • Component • Reliability information for each component • Vulnerability index of each component	Sharma and Trivedi (2007)
Path-based	• Component dependency graph • Interaction among components • Transition probability • Transition reliability	• Sub-domain-based reliability estimation technique • This approach uses dependency among components to analyze them • Operational profiles are assumed to be known • Reliability is defined as the product of transition probability from one component to another, reliability vector on each sub domain of component, and weight vector for each component sub-domain achieved during the transition	• Reliability of components • Weight factor during transition	Zhang et al. (2008)
Path-based	• Modular graph • Transition probability • Failure rate of node	• An adaptive framework incorporating path testing into modular software reliability estimation is proposed • Three techniques used to estimate path reliability: sequence, and branch and loop structures • Path reliability utilized to estimate reliability of the whole application • This method also suggests the effect of failure of each component on the overall reliability of the application	• Number of testing paths in the software system • Number of loops • Number of test cases • Number of detected failures	Hsu and Huang (2011)
Path-based	• Component integration • Component-level reliability • System-level reliability • UML diagrams	• Proposed reliability prediction and estimation technique that helps find the appropriate software components and integrate them in a feasible way • Estimating component properties • Predicting reliability of components • Measuring reliability values of the components provided • Making decisions regarding which components satisfy the required reliability characteristics and are to be selected for integration	• Path-specific reliability value • Number of use cases • Component usage profiles	Palviainen et al. (2011)

3.8 Quality Issues

Quality is one of the most vital properties of any software construct. Quality is related to almost every other attribute of the software including better reusability, reduced cost, usability in the overall software, testability, maintainability, ease of use and reliability. Quality issues should therefore be properly addressed and resolved. The quality of individual components and ultimately of the component-based software is equally crucial.

Major issues related to the quality of components and component-based software are:

- **Identifying and defining quality attributes:** This is a broad area with a good deal of scope. Though many quality attributes are already defined in the literature, most of them are in the context of traditional software or for small-scale applications. Quality attributes that can contribute to the quality of overall component-based software require more focus.

- **Defining quality parameters and procedures for components and component-based software:** Procedures and parameters need to be defined for developing and assessing quality components for complex and large-scale software. Available parameters and procedures address only a small number of quality parameters.

- **Measures and metrics to assess the quality of components and component-based software:** Quality can be defined in many ways and can be assessed through many measures. Component-based software requires specific approaches. Metrics that can predict or assess the quality of components as well as the complete system are decisive.

- **Measures to improve the quality of existing components:** Quality is important for existing as well as newly developed software constructs. The issue is how existing constructs can be reused without compromising the overall quality of the component-based software. Maximum reusability should be achieved without compromising on the quality of the software.

- **Predicting the overall cost and quality of component-based software products:** Practitioners have paid comparatively less attention to exploring the hidden attributes of component-based software. Overall cost estimation techniques can be developed in this domain.

- **Measures to identify level and degree of componentization:** There are no established measures and metrics to find the optimized minimum cost of componentization. The trade-off between the number of components and the integration cost is not yet defined.

- **Maintaining the quality of the component repository:** Newly developed as well as modified components are deposited in the component repository for future use in different applications and in varied contexts. Components provided/purchased by third parties are also placed in the repository. Managing component repositories involves various considerations, including:
  - Size of the repository
  - Cost of maintaining the repository
  - Maintenance effort
  - Selection criteria for components from the repository

- Selection procedure for components available in the repository
- Maintaining versions of the same component in the repository
- Elimination of outdated components from the repository
- Inclusion methods for new components in the repository.

- **Certification procedures for validity and usability of components:** There are no standard procedures to certify the validity of components in component-based software, especially for third-party components. There is no defined method of verifying components' usability other than the customer's requirement specification.

Summary

This chapter discusses critical issues in the development phases of components and component-based software, including:

- Reusability
- Integration and interaction
- Complexity
- Designing and coding
- Testing
- Reliability
- Quality.

This chapter includes a critical literature review of these issues, examining in detail the key areas of work for researchers and academics.

References

Abreu, F. B. 1995. "Design Metrics for Object-Oriented Software System." In *Proceedings Workshop on Quantitative Methods (COOP)*, Aarhus, Denmark, August 1995, 1–30.

Abreu, F. B. and R. Carapuca. 1994. "Object-Oriented Software Engineering: Measuring and Controlling the Development Process." In *4th International Conference on Software Quality*, McLean, VA, 3–5.

Abreu, F. B. and W. Melo. 1996. "Evaluating the Impact of Object-Oriented Design on Software Quality." In *3rd International Software Metrics Symposium*, Berlin, Germany.

Albrecht, A. and J. E. Gaffney. 1983. "Software Function Source Line of Code and Development Effort Prediction: A Software Science Validation." *IEEE Transaction on Software Engineering, SE-9*, 639–648.

Barnard, J. 1998. "A New Reusability Metric for Object-Oriented Software." *Software Quality Journal*, 7: 35–50.

Basili, V. R. and D. H. Hutchens. 1983. "An Empirical Study of a Syntactic Complexity Family." *IEEE Transactions on Software Engineering*, 9(6): 664–672.

Bhattacharya, S. and D. E. Perry. 2005. "Contextual Reusability Metrics for Event-Based Architectures." In *International Symposium on Empirical Software Engineering*, Queensland, Australia, 459–468.

Bixin, L., Y. Zhou, Y. Wang, and J. Mo. 2005. "Matrix Based Component Dependence Representation and Its Applications in Software Quality Assurance." *ACM SIGPLAN Notices*, 40(11): 29–36.

Boehm, B. 1996. "Anchoring the Software Process." *IEEE Software*, 13(4): 73–82.

Boxall, M. and S. Araban. 2004. "Interface Metrics for Reusability Analysis of Components." In *Proceedings of Australian Software Engineering Conference (ASWEC'2004)*, Melbourne, Australia, 40–46.

Caldiera, G. and V. R. Basili. 1991. "Identifying and Qualifying Reusable Software Components." *IEEE Computer*, 24: 61–70.

Chen, J. 2011. "Complexity Metrics for Component-Based Software Systems." *International Journal of Digital Content Technology and Its Applications*, 5(3): 235–244.

Chen, D.-J. and P. J. Lee. 1993. "On the Study of Software Reuse Using Reusable C++ Components." *Journal of Systems Software*, 20(1): 19–36.

Cheung, R. A. 1980. "User-Oriented Software Reliability Model." *IEEE Transaction on Software Engineering*, 6(2): 118–125.

Chidamber, S. and C. Kemerer. 1994. "A Metrics Suite for Object-Oriented Design." *IEEE Transactions on Software Engineering*, 20(6): 476–493.

Cho, E. S., M. S. Kim, and S. D. Kim. 2001. "Component Metrics to Measure Component Quality." In *Proceedings of Eighth Asia-Pacific on Software Engineering Conference (APSEC '01)*, IEEE Computer Society, Washington, DC, 419–426.

Cortellessa, V., H. Singh, and B. Cukic. 2002. "Early reliability assessment of UML based software models." In *Proceedings of 3rd International Workshop on Software and Performance*, Rome.

Dallal, J. and P. Sorenson. 2006. "Generating Class-Based Test Cases for Interface Classes of Object-Oriented Black Box Frameworks." *World Academy of Science, Engineering and Technology*, 16: 96–102, ISSN: 1307-6884.

Edwards, H. S. 2000. "Black-Box Testing Using Flowgraphs: An Experimental Assessment of Effectiveness and Automation Potential." *Software Testing, Verification and Reliability*, 10(4): 249–262.

Elberzhager, F., A. Rosbach, J. Münch, and R. Eschbach. 2012. "Reducing Testing Effort: A Systematic Mapping Study on Existing Approaches." *Information and Software Technology*, 54(10): 1092–1106.

Eusgeld, I., F. C. Freiling, and R. Reussner. 2008. *Dependability Metrics*. LNCS 4909. Springer-Verlag Berlin/Heidelberg, 104–125.

Everett, W. 1999. "Software component reliability analysis." In *Proceedings of IEEE Symposium. Application-Specific Systems and Software Engineering & Technology (ASSET'99)*, Richardson, Texas, 204–211.

Farr, W. 1994. "Software Reliability Modeling Survey." In Lyu M. R. (ed.) *Handbook of Software Reliability Engineering*. McGraw-Hill, New York, 71–117.

Fiondella, L., S. Rajasekaran, and S. Gokhale. 2013. "Efficient Software Reliability Analysis with Correlated Component Failures." *IEEE Transaction on Reliability*, 62: 244–255.

Fraser, G. and A. Arcuri. 2013. "Whole Test Suite Generation." *IEEE Transactions on Software Engineering*, 39(2): 276–291.

Gao, J. Z., H. S. Tsao, and Y. Wu. 2003. *Testing and Quality Assurance for Component-Based Software*. Artech House, Boston, MA.

Gill, N. S. and Balkishan. 2008. "Dependency and Interaction Oriented Complexity Metrics of Component-Based Systems." *ACM SIGSOFT Software Engineering Notes*, 33(2): 1–5.

Gill, N. S. and P. Tomar. 2007. "CBS Testing Requirements and Test Case Process Documentation Revisited." *ACM Sigsoft, Software Engineering Note*, 32(2): 1–4.

Goel, A. L. 1985. "Software Reliability Models: Assumptions, Limitations and Applicability." *IEEE Transaction on Software Engineering*, 11(12): 1411–1423.

Gokhale, S. et al. 1998. "Reliability Simulation of Component Based Software Systems." In *Proceedings of 9th International Symposium on Software Reliability Engineering (ISSRE 98)*, Paderborn, Germany, November 1998, 192–201.

Gokhale, S. S. and K. S. Trivedi. 2006. "Analytical Models for Architecture-Based Software Reliability Prediction: A Unification Framework." *IEEE Transactions on Reliability*, 55(4): 578–590.

Goseva-Popstojanova, K. and K. Trivedi. 2001. "Architecture-Based Approach to Reliability Assessment of Software Systems." *Performance Evaluation an International Journal*, 45: 179–204.

Gui, G. and P. D. Scott. 2008. "New Coupling and Cohesion Metrics for Evaluation of Software Component Reusability." *Proceedings of the International Conference for Young Computer Scientists*, Huan, China, 1181–1186.

Gui, G. and P. D. Scott. 2007. "Ranking Reusability of Software Components Using Coupling Metrics." *The Journal of Systems and Software*, 80: 1450–1459.

Halstead, M. H. 1977. *Elements of Software Science*. Elsevier North Holland, New York.

Henderson-Sellers, B. and D. Tegarden. 1993. "The Application of Cyclomatic Complexity to Multiple Entry/Exit Modules." Center for Information Technology Research Report No. 60.

Henry, S. and D. Kafura. 1981. "Software Structure Metrics Based on Information Flow." *IEEE Transactions on Software Engineering*, 7: 510–518.

Hervé, C., L. Mariani, and M. Pezzè. 2013. "Exception Handlers for Healing Component-Based Systems." *ACM Transactions on Software Engineering and Methodology*, 22(4): 1–40.

Hislop, G. W. 1993. "Using Existing Software in a Software Reuse Initiative." In *Sixth Annual Workshop on Software Reuse (WISR'93)*, Owego, NY, November 2–4, 1993.

Hristov, D., H. Oliver, M. Huq, and W. Janjic. 2012. "Structuring Software Reusability Metrics for Component-Based Software Development." In *Proceedings of Seventh International Conference on Software Engineering Advances ICSEA 2012*, IARIA, Lisbon, Portugal, ISBN: 978-1-61208-230-1.

Hsu, C. J. and C. Y. Huang. 2011. "An Adaptive Reliability Analysis Using Path Testing for Complex Component Based Software Systems." *IEEE Transaction on Reliability*, 60(1): 158–170.

Jain, R., A. Chandrasekaran, G. Elias, and R. Cloutier. 2008. "Exploring the Impact of Systems Architecture and Systems Requirements on Systems Integration Complexity." *IEEE Systems Journal*, 2(2): 209–223.

Poulin, J. S. 1994. "Measuring Software Reusability." In *Proceedings of Third International Conference on Software*, Rio de Janerio, Brazil, November 1–4, 1994, 126–138.

Kharb, L. and R. Singh. 2008. "Complexity Metrics for Component-Oriented Software Systems." *ACM SIGSOFT Software Engineering Notes*, 33(2): 1.

Krishnamurthy, S. and A. P. Mathur. 1997. "On the Estimation of Reliability of a Software System Using Reliabilities of Its Components." In *Proceedings of 8th International Symposium on Software Reliability Engineering (ISSRE'97)*, Albuquerque, NM, 146–155.

Kubat, P. 1989. "Assessing Reliability of Modular Software." *Operations Research Letters*, 8: 35–41.

Kumari, U. and S. Bhasin. 2011. "A Composite Complexity Measure for Component-Based Systems." *ACM SIGSOFT Software Engineering Notes*, 36(6): 1–5.

Lau, K.-K. and Z. Wang. 2007. "Software Component Models." *IEEE Transactions on Software Engineering*, 33(10): 709–724.

Lee, Y. and K. H. Chang. 2000. "Reusability and Maintainability Metrics for Object-Oriented Software." In *Proceedings of 38th Annual on Southeast Regional Conference (ACM-SE 38)*. ACM, New York, 88–94.

Littlewood, B. 1975. "A Reliability Model for Systems with Markov Structure." *Applied Statistics*, 24(2): 172–177.

Littlewood, B. B. and L. Strigini. 1993. "Validation of Ultra-High Dependability for Software-Based Systems." *Communications of the ACM*, 36(11): 69–80.

McCabe, T. 1976. "A Complexity Measure." *IEEE Transactions on Software Engineering*, 2(8): 308–320.

Michael, L. R., R. Sampath, and P. A. Aad van Moorsel. 2002. "Optimal Allocation of Test Resources for Software Reliability Growth Modeling in Software Development." *IEEE Transactions on Reliability*, 51(2): 183–192.

Morris, K. 1989. "Metrics for Object Oriented Software Development." Masters thesis, M.I.T., Sloan School of Management, Cambridge, MA.

Myers, G. J. 2004. *The Art of Software Testing*, 2nd edn. John Wiley & Sons, Hoboken, NJ.

Musa, J. 1998. *Software Reliability Engineering*. McGraw-Hill, New York.

Narasimhan, V. L. and B. Hendradjaya. 2005. "Theoretical Considerations for Software Component Metrics." *Transactions on Engineering, Computing and Technology*, 10: 169–174.

Naur, P. and B. Randall. 1969. *Software Engineering: A Report on a Conference Sponsored by the NATO Science Committee*. Scientific Affairs Division, NATO, Brussels.

Ntafos, S. C. 1988. "A Comparison of Some Structural Testing Strategies." *IEEE Transactions in Software Engineering*, 14(6): 868–874.

Orso, A., M. J. Harrold, D. Rosenblum, G. Rothermel, M. L. Soffa, and H. Do. 2001. "Using Component Meta Contents to Support the Regression Testing of Component-Based Software." In *Proceedings of International Conference on Software Maintenance*, Florence, Italy, 716–725.

Ostrand, T. J. and M. J. Balcer. 1988. "The Category-Partition Method for Specifying and Generating Functional Tests." *Communications of the ACM*, 31(6): 676–686.

Palviainen, M., A. Evesti, and E. Ovaska. 2011. "The Reliability Estimation, Prediction and Measuring of Component-Based Software." *Journal of System and Software*, 84: 1054–1070.

Poulin, J. S. 1996. *Measuring Software Reuse–Principles, Practices and Economic Models*. Addison-Wesley, Reading, MA.

Pressman S. R. 2005. *Software Engineering A practitioners Approach*, 6th. edn TMH International Edition, New York.

Prieto-Diaz, R. and P. Freeman. 1987. "Classifying Software for Reusability." *IEEE Software*, 4(1): 6–16.

Ramamoorthy, C. V. and F. B. Bastani. 1982. "Software Reliability-Status and Perspectives." *IEEE Transactions on Software Engineering*, 8(4): 354–371.

Ramamoorthy, C., S. Ho, and W. Chen. 1976. "On the Automated Generation of Program Test Data." *IEEE Transactions on Software Engineering*, 2(4): 293–300.

Reussner, R. H., H. W. Schmidt, and I. H. Poernomo. 2003. "Reliability Prediction for Component-Based Software Architectures." *Journal Systems Software*, 66(3): 241–252.

Rodrigues, G. N., D. S. Rosenblum, and S. Uchitel. 2005. "Using Scenarios to Predict the Reliability of Concurrent Component-Based Software systems." In *Proceedings of 8th International Conference on Fundamental Approaches to Software Engineering, FASE 2005*, Springer Lecture Notes in Computer Science, Edinburgh.

Sanyal, S. et al. 2004. "Framework of a Software Reliability Engineering Tool." In *Proceedings of IEEE High-Assurance Systems Engineering Workshop (HASE'97)*, Washington, DC, 114–119.

Sen, T. and R. Mall. 2012. "State-Model-Based Regression Test Reduction for Component-Based Software." *International Scholarly Research Network ISRN Software Engineering*, 2012: 1–11.

Sharma, A. and D. S. Kushwaha. 2012. "Applying Requirement Based Complexity for the Estimation of Software Development and Testing Effort." *ACM SIGSOFT Software Engineering Notes*, 37(1): 1–11.

Sharma, V. S. and S. K. Trivedi. 2007. "Quantifying Software Performance, Reliability and Security: An Architecture-Based Approach." *The Journal of Systems and Software*, 80: 493–509.

Shooman, M. 1997. "Structural Models for Software Reliability Prediction." In *Proceedings of 2nd International Conference on Software Engineering*, Providence, RI, 268–280.

Singh, H., B. Cortellessa, E. G. Cukic, and V. Bharadwaj. 2001. "A Bayesian Approach to Reliability Prediction and Assessment of Component Based Systems." In *Proceedings of 12th IEEE International Symposium on Software Reliability Engineering*, Hong Kong, 12–21.

Tiwari, U. K. and S. Kumar. 2014. "In-Out Interaction Complexity Metrics for Component-Based Software." *ACM SIGSOFT Software Engineering Notes*, 39(5): 1–4.

Tiwari, U. K. and S. Kumar. 2016. "Components Integration-Effect Graph: A Black Box Testing and Test Case Generation Technique for Component-Based Software." *International Journal of Systems Assurance Engineering and Management*, 8(2): 393–407.

Trevor, P., A. Mos, M. Trofin, T. Gschwind, and J. Murphy. 2008. "Extracting Interactions in Component-Based Systems." *IEEE Transactions on Software Engineering*, 34(6): 783–799.

Tsai, W. T., A. Saimi, L. Yu, and R. Paul. 2003. "Scenario-Based Object-Oriented Testing Framework." In *Proceedings of Third International Conference on Quality Software (QSIC'03), IEEE*, Dallas, TX, 410–417.

Vitharana, P., H. Jain, and F. Zahedi. 2004. "Strategy-Based Design of Reusable Business Components." *IEEE Transactions on Systems, Man, and Cybernetics—PART C: Applications and Reviews*, 34(4): 460–474.

Voas, J. M. 1992. "A Dynamic Testing Complexity Metric." *Software Quality Journal*, 1(2): 101–114.

Voas J. M. and K. W. Miller. 1995. "Software Testability: The New Verification." *IEEE Software*, 12(3): 17.

Voas, J. M. and K. W. Miller. 1992. "The Revealing Power of a Test Case." *Journal of Software Testing, Verification and Reliability*, 2(1): 25–42.

Wake, S. and S. Henry. 1988. "A Model Based on Software Quality Factors which Predict Maintainability." In *Proceedings of Conference on Software Maintenance*, Phoenix, Arizona, 382–387.

Washizaki, H., Y. Hirokazu, and F. Yoshiaki. 2003. "A Metrics Suite for Measuring Reusability of Software Components." In *Proceedings of 9th International Symposium on Software Metric*, Sydney, Australia, 211–223m

Weyuker, E. J. 1993. "More Experience with Data Flow Testing." *IEEE Transactions on Software Engineering*, 19(9): 912–919.

Weyuker, E. J. 1998. "Testing Component-Based Software: A Cautionary Tale." *IEEE Software*, 15(5): 54–59.

Wijayasiriwardhane, T. and R. Lai. 2010. "Component Point: A System-Level Size Measure for Component-Based Software Systems." *The Journal of Systems and Software*, 83: 2456–2470.

Xie, M. and C. Wohlin. 1995. "An Additive Reliability Model for the Analysis of Modular Software Failure Data." In *Proceedings of 6th International Symposium on Software Reliability Engineering (ISSRE'95)*, Toulouse, France, 188–194.

Yacoub, S., B. Cukic, and H. Ammar. 2004. "A Scenario-Based Reliability Analysis Approach for Component Based Software." *IEEE Transactions on Reliability*, 53(4): 465–480.

Zhang, F., X. Zhou, J. Chen, and Y. Dong. 2008. "A Novel Model for Component–based Software Reliability Analysis." In *Proceedings of 11th IEEE High Assurance Systems Engineering Symposium*, Nanjing, China, 303–309.

4

Reusability Metric-Based Selection and Verification of Components

4.1 Introduction

Effective and well-organized reuse of software deliverables, work products and artifacts offer a return on the investment in terms of a shorter development cycle, effective cost, improved quality and increased productivity. Software reusability is the effective reuse of pre-designed and tested parts of experienced software in new applications. This chapter presents a reusability metric to estimate the reusability of components in component-based software engineering. To estimate the reusability metric of the different classes of components, function points (Albrecht and Gaffney 1983) are used as the base metric defining the "reusability matrix" that identifies the component reusability ratio. We can use lines of code to compute the reusability metric, where function points are not available for the component.

4.2 Reuse and Reusability

Researchers and practitioners have given an array of definitions for the term "software reuse." Software reuse is the process of constructing software from existing software artifacts rather than developing it from scratch. Krueger defined software reuse as "the process of creating software systems from existing software rather than building them from scratch" (Krueger 1992). Software reuse is the process of integrating predefined specifications, design architectures, tested code, or test plans with the proposed software (Bersoff and Davis 1991, Johnson and Harri 1991, Maiden and Sutcliff 1991). According to Freeman, "reuse is the use of any information which a developer may need in the creation process" (Freeman 1987). Basili and Rombach (1988) suggest reuse as the use of existing knowledge of any type or category mapped with the software project under development. In the Tracz view, reuse means the use of software systems that were developed in the past for reuse (Tracz 1995). Braun considers reuse in terms of use of existing components in the same context or in a context-free environment (Braun 1994).

A number of eminent researchers have defined reuse in terms of adaptability or modifiability of the reused artifact. Lim proposes that the work products and deliverables of the previous software are to be used in new software development without any modification (Lim 1994). On the other hand, McIlroy defines reuse as the use of off-the-shelf components with the ability to modify but in a managed and controlled manner (McIlroy 1976). Cooper describes software reuse as the potential repeated use of a formerly designed software component, with minor, major or no modifications (Cooper 1994). Software reusability is the level of usefulness or the extent of reuse. Reusability is defined as the amount of ease with which pre-designed objects can be used in a new context (Biggerstaff and Richter 1989, Prieto-Diaz 1992, Kim and Stohr 1998).

In literature, software reuse has been defined as a concept or a process, whereas software reusability is defined as the outcome of that process. Reusability is the actual level of implementation of reuse. Reusability acts as a measurement tool to quantify the proportion of reuse of a pre-designed, implemented and qualified artifact. The present work defines a quantifiable approach to measure the reusability of a component in component-based software.

4.3 Reusable Artifacts and Work Products

Any information or knowledge acquired from previous or existing software that can be used in the development of the new software is known as a reusable artifact. Freeman defined a reusable work product as an object of reusability which a developer needs in the construction of a new system (Freeman 1987). Reusable artifacts can be a part of the previous software or this may be used in its entirety.

A reusable artifact may be any of the following software deliverables:

- *Domain knowledge:* facts and rules of the domain that are uniformly applicable to all the solutions of that particular domain.
- *Development process:* steps and methodology reused to construct new software through which previous software was implemented successfully.
- *Conceptual design and architecture:* either in the form of theoretical documents or the implementation part of the previous application.
- *Requirements specification:* defined for earlier software and fits a new application, a part or the whole document. Specifications include both functional and non-functional requirements.
- *Design documents:* include graphical interfaces as well as database designs. Design includes system-level design and detailed design of each component.
- *Code segments and code lists:* pre-compiled and pre-executed code fragments developed for previous software. It includes the code of functions, modules or data structures.
- *Interfaces:* used to interact and integrate various components without compatibility- or context-related problems.
- *Test cases and test plans:* developed for and applied to experienced applications and may be reused in a new context.
- *Test data:* used as an input in testing of software components.
- *User feedbacks:* feedback from earlier software can be an important factor in developing new software fulfilling the needs of the user.

Only some artifacts are included here, though there is an enormous number of reusable software artifacts that can be identified. Reusable work products have certain properties that they mutually contribute to explore and support reusability. These characteristics include but are not limited to (Biggerstaff and Richter 1989, Matsumoto 1989, McClure 1989):

- *Applicability and expressiveness* in different contexts.
- *Unambiguously defined objectives* with their abilities and disabilities.
- *Context-free portability* so that it can be implemented without parametric problems.
- *Adaptability* so that components can be reused by making modifications without introducing new errors.
- *Well-defined and documented.*
- *Self-descriptive* containing necessary information regarding component behavior.

Software artifacts can be characterized at various levels, including statement level, function level, component level or system level (Maxim 2019).

a. *Statement-level* reusability defines the reuse of single-line statements, such as predefined syntactical statements or semantics of syntax.

b. *Function-level* reusable artifacts are the implementation of single well-defined functions.

c. *Component-level* reuse describes sub-routines or complete objects of previously developed software reused in a new application.

d. *System-level reusability* is about reusing an entire application.

4.4 Reusability Metrics

In component-based software development, reused and new components are assembled together to develop component-based applications. The comparison between the reused and the new components defines the metric of component reusability. In this chapter, we define a quantifiable approach to measuring the reusability of a component in component-based software. Using the level and degree of reusability, we categorize components into four classes: newly developed, fully qualified adaptable, partially qualified adaptable and off-the-shelf components (Brown and Wallnau 1998, Atkinson et al. 2002).

a. **Newly developed components:** These are constructed from scratch according to the requirements of an individual application. These components are developed either by the development team or by some third party with experience in developing reusable components.

b. **Partially qualified adaptable components:** These need a major degree of alteration in their code to fit into the software under development.

c. **Fully qualified adaptable components:** These require a minor degree of adaptation to qualify for the needs of the current application. Their reusability level is much higher than that of partially qualified components.

d. **Off-the-shelf components:** These are pre-designed, pre-constructed and pre-tested components. They are third-party components or are built up by the development team as part of the earlier project and can be reused in the current software without any modification or change.

In component-based software development, reused and new components are assembled together to develop component-based applications. The comparison between the reused and the new components defines the component reusability metric.

Function points are the base metric used to define the reusability metric. Alan Albrecht (Albrecht and Gaffney 1983) proposed the function-point technique to measure complexity in terms of size and functionalities provided by the system. This method addresses the functionality provided by the system from the user's point of view. To analyze the software system, Albrecht divided it into five functional units on the basis of data consumed and produced. Three complexity weights—high, low and medium—are associated with these functional units, using a set of predefined values. In function-point analysis (FPA), 14 complexity factors have been defined, with a rating from 0 to 5. On the basis of these factors, Albrecht calculated the unadjusted function-point value, the complexity adjustment factors and finally the value of the function points.

FPA categorizes all the functionalities provided by the software into five specific functional units:

1. **External inputs** are the number of distinct data inputs provided to the software or the control information inputs that modify the data in internal logical files. The same inputs provided with the same logic are not included in the count for every occurrence. All the repeated formats are treated as one count.

2. **External outputs** are the number of distinct data or control outputs that are provided by the software. The same outputs achieved with the same logic are not included in the count for every occurrence. All the repeated formats are treated as one count.

3. **External inquiries** are the number of inputs or outputs provided to or achieved from the system under consideration without making any change in the internal logical files. The same inputs/outputs with the same logic are not included in the count for every occurrence. All the repeated formats are treated as one count.

4. **Internal logical files** present the number of user data and content residing in the system or control information produced or used in the application.

5. **External interface files** are the number of communal data, contents, files or control information that are accessed, provided or shared among the various applications of the system.

These five functional units are categorized into three levels of complexity: low/simple, average/medium or high/complex. Albrecht identified and defined weights for these complexities with respect to all five functional units. The functional units and their corresponding weights are used to count the unadjusted function points:

$$\text{Unadjusted FP} = \sum_{i=1}^{5}\sum_{j=1}^{3}\left(\text{Count of Functional Unit} * \text{Weight of the Unit}\right)$$

where i denotes the five functional units and j denotes the level of complexity.

Similarly, Albrecht defined complexity adjustment factors on the basis of 14 complexity factors on a scale of 0–5. Adjustment factors provide an adjustment of ±35% ranging from 0.65 to 1.35. The complexity factors are reliable backup and recovery, communication requirement, distributed processing, critical performance, operational environment, online data entry, multiple screen inputs, master file updates, complex functional units, complex internal processing, reused code, conversions, distributed installations and ease of use. Complexity factors are rated as no influence (0), incidental (1), moderate (2), average (3), significant (4) and essential (5).

The complexity adjustment factor is defined as

$$\left[0.65 + 0.01 * \sum_{k=1}^{14} \left(\text{Complexity Factor} \right)_i \right]$$

The function point is defined as the product of unadjusted FP and the complexity adjustment factor.

$$FP = \text{Unadjusted FP} * \text{Complexity adjustment factor}$$

The function-point technique does not depend on tools, technologies or languages used to develop the program or software. Two dissimilar programs having different lines of code may provide the same number of function points. These estimates are not based on lines of code, hence estimations can be made early in the development phase, even after the commencement of the requirements phase.

All these metrics can be defined with the help of lines of code (LOC).

4.4.1 Common Acronyms and Notations

C_i	Component "i"
C_{CBS}	Total number of components in the component-based software
$|C_{New}|$	Cardinality of new components
$|C_{Reused}|$	Cardinality of reused components selected from the repository
$|C_{Off-the-shelf}|$	Off-the-shelf components
$C_{Adaptable}$	Adaptable/ modifiable components
$C_{Full-qual}$	Fully qualified adaptable components
$C_{Part-qual}$	Partially qualified components
RMC_i	Reusability metric of component C_i
RMC_{CBS}	Reusability metric of component-based software
$RMC_{Full-Qual}$	Reusability metric of fully qualified components
$RMC_{CBS-Full-Qual}$	Reusability metric of component-based software at system level
$RMC_{i-Full-Qual}$	Reusability metric of component-based software at component level
$RMC_{Part-Qual}$	Reusability metric of partially qualified component
$RMC_{CBS-Part-Qual}$	Reusability metric of partially qualified component at system level
$RMC_{i-Part-Qual}$	Reusability metric of partially qualified component at component level
$RMC_{Off-the-shelf}$	Reusability metric of off-the-shelf components
$RMC_{CBS-Off-the-shelf}$	Reusability metric of off-the-shelf components at application level
$RMC_{i-Off-the-shelf}$	Reusability metric of off-the-shelf components at component level

4.4.2 Function-Point (FP) Acronyms and Notations

$\lvert \text{FPC}_{\text{CBS}} \rvert$	Total function points of component-based software
$\lvert \text{FPC}_i \rvert$	Total function points of component C_i
$\lvert \text{FPC}_{\text{CBS-Full-qual}} \rvert$	Total number of fully qualified function points of component-based software
$\lvert \text{FPC}_{i\text{-Full-qual}} \rvert$	Total number of fully qualified function points of component C_i
$\lvert \text{FPC}_{\text{CBS-Reused}} \rvert$	Total reused function points of component-based software
$\lvert \text{FPC}_{i\text{-Reused}} \rvert$	Total number of reused function points of component C_i
$\lvert \text{FPC}_{\text{CBS-Adaptable}} \rvert$	Total number of adaptable function points of component-based software
$\lvert \text{FPC}_{i\text{-Adaptable}} \rvert$	Total number of adaptable function points of component C_i
$\lvert \text{FPC}_{\text{CBS-Part-qual}} \rvert$	Total number of partially qualified function points of component-based software
$\lvert \text{FPC}_{i\text{-Part-qual}} \rvert$	Total number of partially qualified function points of component C_i
$\lvert \text{FPC}_{\text{CBS-Off-the-shelf}} \rvert$	Total number of off-the-shelf function points of component-based software
$\lvert \text{FPC}_{i\text{-Off-the-shelf}} \rvert$	Total number of off-the-shelf function points of component C_i

4.4.3 Lines-of-Code (LOC) Acronyms and Notations

$\lvert \text{LOCC}_{\text{CBS}} \rvert$	Total lines of code of component-based software
$\lvert \text{LOCC}_i \rvert$	Total lines of code of component C_i
$\lvert \text{LOCC}_{\text{CBS-Full-qual}} \rvert$	Total number of fully qualified lines of code of component-based software
$\lvert \text{LOCC}_{i\text{-Full-qual}} \rvert$	Total number of fully qualified lines of codes of component C_i,
$\lvert \text{LOCC}_{\text{CBS-Reused}} \rvert$	Total reused lines of code of component-based software
$\lvert \text{LOCC}_{i\text{-Reused}} \rvert$	Total number of reused lines of code of component C_i
$\lvert \text{LOCC}_{\text{CBS-Adaptable}} \rvert$	Total number of adaptable lines of codes of component-based software
$\lvert \text{LOCC}_{i\text{-Adaptable}} \rvert$	Total number of adaptable lines of code of component C_i
$\lvert \text{LOCC}_{\text{CBS-Part-qual}} \rvert$	Total number of partially qualified lines of code of component-based software
$\lvert \text{LOCC}_{i\text{-Part-qual}} \rvert$	Total number of partially qualified lines of code of component C_i
$\lvert \text{LOCC}_{\text{CBS-Off-the-shelf}} \rvert$	Total number of off-the-shelf lines of code of component-based software
$\lvert \text{LOCC}_{i\text{-Off-the-shelf}} \rvert$	Total number of off-the-shelf lines of code of component C_i

For the reusability metric we define four types of function points of a component:

a. Total function points

b. Reused function points

c. Adaptable function points

d. New function points.

a. Total Function Points of a Component

Using Function Points

These are defined as the summation of reused function points and new function points:

$$\lvert \text{FPC}_i \rvert = \lvert \text{FPC}_{i-\text{Reused}} \rvert + \lvert \text{FPC}_{\text{New}} \rvert \tag{4.1}$$

where FPC_i is the total number of function points of a component, $\text{FPC}_{i-\text{Reused}}$ is the count of reusable function points and FPC_{New} is the number of new function points.

Using Lines of Code (LOC)

On the basis of lines of code (LOC), we can define the metric as: $\lvert \text{LOCC}_i \rvert = \lvert \text{LOCC}_{i-\text{Reused}} \rvert + \lvert \text{LOCC}_{\text{New}} \rvert$

b. **Reused Function Points of a Component**
Using Function Points
This includes off-the-shelf function points and adaptable function points of a component.

$$C_{i-Reused} = \left| FPC_{i-Off-the-shelf} \right| + \left| FPC_{i-Full-qual} \right| + \left| FPC_{i-Part-qual} \right| \tag{4.2}$$

where $FPC_{i-Reused}$ is the count of reusable function points, $FPC_{i-Off-the-shelf}$ is the count of off-the-shelf function points, $FPC_{i-Full-qual}$ is the number of fully qualified function points and $FPC_{i-Part-qual}$ is the number of partially qualified function points.

Using Lines of Code (LOC)
On the basis of lines of code (LOC), we can define the metric as:

$$LOCC_{i-Reused} = \left| LOCC_{i-Off-the-shelf} \right| + \left| LOCC_{i-Full-qual} \right| + \left| LOCC_{i-Part-qual} \right|$$

where $LOCC_{i-Reused}$ is the reused lines of code, $LOCC_{i-LOCCi-Off-the-shelf}$ is the number of lines of code which can be reused as they are, $LOCC_{i-Full-qual}$ represents the fully qualified lines of code and $LOCC_{i-Part-qual}$ is the number of lines of code that are partially qualified.

c. **Adaptable Function Points of a Component**
Using Function Points
Adaptable components include partially qualified as well as fully qualified components.

$$C_{i-Adaptable} = \left| FPC_{i-Full-qual} \right| + \left| FPC_{i-Part-qual} \right| \tag{4.3}$$

where $C_{i-Adaptable}$ is the total number of reusable function points, $FPC_{i-Full-qual}$ is the number of fully qualified function points and $FPC_{i-Part-qual}$ is the number of partially qualified function points.

Using Lines of Code (LOC)
On the basis of lines of code (LOC), we can define the metric as:

$$LOCC_{i-Adaptable} = \left| LOCC_{i-Full-qual} \right| + \left| LOCC_{i-Part-qual} \right|$$

where $LOCC_{i-Adaptable}$ is the total number of reusable lines of code, $LOCC_{i-Full-qual}$ is the number of fully qualified lines of code and $LOCC_{i-Part-qual}$ is the number of partially qualified lines of code.

d. **New Function Points of a Component**
Using Function Points
These are the function points which are achieved by excluding reused function points from the total function points of the component.

$$C_{New} = \left| FPC_i \right| - \left| FPC_{i-Reused} \right| \tag{4.4}$$

where C_{New} denotes the new function points, $FPC_{i-Reused}$ is the count of reusable function points.

Using Lines of Code (LOC)
On the basis of lines of code (LOC), we can define the metric as:

$$LOCC_{New} = |LOCC_i| - |LOCC_{i-Reused}|$$

where C_{New} represents the new lines of code and $LOCC_{i-Reused}$ is the reusable line of code.

In this chapter, two levels of reusability metrics are proposed—*component level* and *CBS system level*.

a. Component-level Reusability Metric (RMC$_i$)

Using Function Points
Using function points, the reusability metric at the component level is defined as the ratio between the total number of reused function points of the component and the total number of function points of that component.

$$RMC_i = \frac{|FPC_{i-Reused}|}{|FPC_i|} \tag{4.5}$$

where RMC_i denotes the reusability metric at the component level, $FPC_{i-Reused}$ denotes the reusable function points and FPC_i denotes the total number of function points of the component.

Using Lines of Code (LOC)
On the basis of lines of code (LOC), we can define the metric as:

$$RMC_i = \frac{|LOCC_{i-Reused}|}{|LOCC_i|}$$

where RMC_i denotes the reusability metric at component level, $LOCC_{i-Reused}$ denotes the reusable lines of code and $LOCC_i$ denotes the total number of LOCs of the component.

b. CBS System-level Reusability Metric (RMC$_{CBS}$)

At CBS-system level, the reusability metric is defined in two ways: in terms of cardinality of components, and in terms of function points.

Reusability Metric in Terms of Cardinality of Components

Using Function Points
In terms of components, the reusability metric is computed by considering the number of new, partially qualified, fully qualified and off-the-shelf components. It is achieved by taking the ratio between the number of reused components and the total number of components involved in the component-based application.

$$RMC_{CBS} = \frac{|C_{Reused}|}{C_{CBS}} \tag{4.6}$$

Using Lines of Code (LOC)
On the basis of lines of code (LOC), we can define the metric as:

$$RMC_{CBS} = \frac{|LOC_{Reused}|}{|LOC_{CBS}|}$$

where RM_{Reused} is the reusability metric at the component-based software, LOC_{Reused} denotes the count of reused lines of code and LOC_{CBS} denotes the total number of LOCs at the component-based software level.

Reusability Metric in Terms of Function Points

Using Function Points
Here, the reusability metric is computed by counting the number of new, partially qualified, fully qualified and off-the-shelf function points of the whole component-based software. When function points are available, the reusability metric is defined as the ratio between the total reused function points of all the components and the total function points of the CBS application:

$$RMC_{CBS} = \frac{\left|FPC_{CBS-Reused}\right|}{\left|FPC_{CBS}\right|} \qquad (4.7)$$

Using Lines of Code (LOC)
On the basis of lines of Code (LOC), we can define the metric

$$RMC_{CBS} = \frac{\left|LOCC_{CBS-Reused}\right|}{\left|LOCC_{CBS}\right|}$$

where RMC_{CBS} is the reusability metric at the component-based application level, $LOCC_{CBS-Reused}$ denotes the count of reused lines of code and $LOCC_{CBS}$ denotes the total number of LOCs at the application level.

In the present work, reusability metrics for the following categories of component are defined:

- *Adaptable components*, which includes *fully qualified* adaptable and *partially qualified* adaptable components
- *Off-the-shelf components*, which can be reused without any modification.

4.4.4 Adaptable Reusability Metric (RM$_{Adaptable}$)

Adaptable components are those components that can accommodate existing requirements, design, code or test cases with minor or major modifications. Adaptable components are divided into two categories according to their degree of modification:

A. *Fully qualified components*
B. *Partially qualified components*

Fully qualified components require no modification or a minor modification and partially qualified components require major modification. There are two scenarios in each category:

Scenario 1: *Component-level adaptable reusability metric*
Scenario 2: *CBS system-level adaptable reusability metric*

In this chapter, the reusability assessment for both is defined in terms of components as well as of function points.

4.4.4.1 *Fully Qualified Components*

> ### SCENARIO 1: COMPONENT-LEVEL ADAPTABLE REUSABILITY METRIC ($RMC_{i\text{-FULL-QUALIFIED}}$)
>
> Fully qualified components require a little alteration to fit into the new contexts. At component level, reusability is assessed in terms of function points. Three reusability metric cases are described for fully qualified components:
>
> Case 1: Reusability metric when all parts of the component are involved.
> Case 2: Reusability metric when only reused parts of the component are involved.
> Case 3: Reusability metric when only adaptable parts of the component are involved.
>
> Every case can be defined using function points as well lines of code.
>
> i. **When all parts of the component are involved**
>
> *Using function points*
>
> Here the ratio of fully qualified function points of a particular component, C_i, to the total number of function points of that component is defined as:
>
> $$RMC_{i\text{-Full-qualified}} = \frac{\left|FPC_{i\text{-Full-qualified}}\right|}{\left|FPC_i\right|} \qquad (4.8)$$
>
> where FPC_i denotes the total function points of a component including new, adaptable and reused.
>
> *Using lines of code (LOC)*
>
> On the basis of lines of code (LOC), we can define the metric as
>
> $$RMC_{i\text{-Full-qualified}} = \frac{\left|LOCC_{i\text{-Full-qualified}}\right|}{\left|LOCC_i\right|}$$
>
> where $LOCC_i$ denotes the total LOCs of a component including new, customized and reused.
>
> ii. **When only reused parts of the component are involved**
>
> *Using function points*
>
> In this context the ratio is taken with respect to only reused function points of the component. Here the total number of fully qualified function points is divided by the total number of reused function points.
>
> Therefore, the reusability metric in terms of function points is defined as:
>
> $$RMC_{i\text{-Full-qualified}} = \frac{\left|FPC_{i\text{-Full-qualified}}\right|}{\left|FPC_{i\text{-Reused}}\right|} \qquad (4.9)$$
>
> where total reused function points $FPC_{i-Reused}$ represent the collection of off-the-shelf, fully qualified and partially qualified adaptable components from Equation (4.2).

Using lines of code (LOC)

On the basis of lines of codes (LOC), we can define the metric as

$$RMC_{i-Full-qualified} = \frac{\left|LOCC_{i-Full-qualified}\right|}{\left|LOCC_{i-Reused}\right|}$$

where $LOCC_{i-Reused}$ denotes the collection of off-the-shelf, fully qualified and partially qualified reusable LOCs.

iii. **When only adaptable parts of the component are involved**

Using function points

This case takes the ratio of adaptable function points only. The total number of fully qualified function points is divided by the total numbers of adaptable function points of the component to give a reusability metric of:

$$RMC_{i-Full-qualified} = \frac{\left|FPC_{i-Full-qualified}\right|}{\left|FPC_{i-Adaptable}\right|} \qquad (4.10)$$

where adaptable function points $FPC_{i-Adaptable}$ are the assembly of fully qualified and partially qualified function points only.

Using lines of code (LOC)

On the basis of lines of code (LOC), we can define the metric as

$$RMC_{i-Full-qualified} = \frac{\left|LOCC_{i-Full-qualified}\right|}{\left|LOCC_{i-Adaptable}\right|}$$

where adaptable LOCs are the assembly of fully qualified and partially qualified LOCs only.

SCENARIO 2:   CBS SYSTEM-LEVEL ADAPTABLE REUSABILITY METRIC (RMC_{CBS-FULL-QUALIFIED})

There are three different cases of adaptable reusability metric in terms of component count as well as in terms of function points:

Case 1: Reusability metric when all parts of the component are involved.
Case 2: Reusability metric when only reused parts of the component are involved.
Case 3: Reusability metric when only adaptable parts of the component are involved.

Each case can be defined using function points as well as line of code.

i. **When all components are involved**

In terms of component count

In this case, the ratio is taken in the context of the total number of components involved in the CBS system. Here, the total number of fully qualified reused components is divided by the total number of components involved in the CBS development. It is defined as:

$$RMC_{CBS-Full-qualified} = \frac{\left|C_{Full-qualified}\right|}{C_{CBS}} \qquad (4.11)$$

In terms of function points

In this case, we define the reusability metric for fully qualified components as the ratio between the total function points of the component-based software and the fully qualified function points of the CBS:

$$RMC_{CBS-Full-qualified} = \frac{\left|FPC_{CBS-Full-qualified}\right|}{\left|FPC_{CBS}\right|} \qquad (4.12)$$

In terms of lines of code

In this case, we define the reusability metric for fully qualified components as the ratio between the total lines of code of the component-based software and the fully qualified lines of code of the CBS:

$$RMC_{CBS-Full-qualified} = \frac{\left|LOCC_{CBS-Full-qualified}\right|}{\left|LOCC_{CBS}\right|}$$

ii. When only reused components are involved

In terms of component count

In this case, the ratio is taken in the context of reused components only. It is defined as the total number of fully qualified reused components divided by the total number of reused components involved in the CBS application development:

$$RMC_{CBS-Full-qualified} = \frac{\left|C_{Full-qualified}\right|}{\left|C_{Reused}\right|} \qquad (4.13)$$

In terms of function points

In this case, the reusability metric can be given as the ratio between the number of fully qualified function points and the reused function points of the component-based software:

$$RMC_{CBS-Full-qualified} = \frac{\left|FPC_{CBS-Full-qualified}\right|}{\left|FPC_{CBS-Reused}\right|} \qquad (4.14)$$

In terms of lines of code

In this case, the reusability metric can be given as the ratio between the number of fully qualified lines of code and the reused lines of code of the component-based software:

$$RMC_{CBS-Full-qualified} = \frac{\left|LOCC_{CBS-Full-qualified}\right|}{\left|LOCC_{CBS-Reused}\right|}$$

iii. **When only adaptable components are involved**

In terms of component count

In this case, the ratio is taken in the context of adaptable components only. Here, the total number of fully qualified reused components is divided by the total number of adaptable components involved in the development. It is defined as:

$$RMC_{CBS-Full-qualified} = \frac{\left|C_{Full-qualified}\right|}{\left|C_{Adaptable}\right|} \tag{4.15}$$

In terms of function points

Here, the reusability metric is defined as the total fully qualified function points divided by the total number of adaptable function points of the CBS:

$$RMC_{CBS-Full-qualified} = \frac{\left|FPC_{CBS-Full-qualified}\right|}{\left|FPC_{CBS-Adaptable}\right|} \tag{4.16}$$

In terms of lines of code

Here, the reusability metric is defined as the total fully qualified lines of code divided by the total number of adaptable lines of code of the CBS:

$$RMC_{CBS-Full-qualified} = \frac{\left|LOCC_{CBS-Full-qualified}\right|}{\left|LOCC_{CBS-Adaptable}\right|}$$

4.4.4.2 Partially Qualified Components

Two scenarios apply to partially qualified components, as for fully qualified components.

SCENARIO 1:   COMPONENT-LEVEL ADAPTABLE REUSABILITY METRIC ($RMC_{I-PART-QUALIFIED}$)

There are three different contexts for the reusability metric for partially qualified adaptable components at component level:

Case 1: Reusability metric when all parts of the component are involved.
Case 2: Reusability metric when only reused parts of the component are involved.
Case 3: Reusability metric when only adaptable parts of the component are involved.

Each case can be defined using function points as well as line of code.

i. **When all parts of the component are involved**

Using function points

In this case, we calculate the reusability metric in the context of a particular component when all parts of the components are involved. Here, the total number of partially qualified reused function points is divided by the total number of function points of the components. That is:

$$RMC_{i-Part-qualified} = \frac{\left|FPC_{i-Part-qualified}\right|}{\left|FPC_i\right|} \tag{4.17}$$

Using lines of code

Here, the total number of partially qualified reused lines of code is divided by the total number of lines of code of the components. That is:

$$RMC_{i-Part-qualified} = \frac{\left|LOCC_{i-Part-qualified}\right|}{\left|LOCC_i\right|}$$

ii. When only reused parts of the component are involved

Using function points

In this context the ratio only refers to the reused function points of the component. The total number of partially qualified function points is divided by the total number of reused function points. Hence, it is defined as:

$$RMC_{i-Part-qualified} = \frac{\left|FPC_{i-Part-qualified}\right|}{\left|FPC_{i-Reused}\right|} \tag{4.18}$$

Using lines of code (LOC)

In this case the total number of partially qualified lines of code is divided by the total number of reused lines of code. Hence, it is defined as:

$$RMC_{i-Part-qualified} = \frac{\left|LOCC_{i-Part-qualified}\right|}{\left|LOCC_{i-Reused}\right|}$$

iii. When only adaptable parts of the component are involved

Using function points

In this case, the ratio refers to adaptable function points only. The total number of fully qualified function points is divided by the total number of adaptable function points only. Therefore, the reusability metric is defined as:

$$RMC_{i-Part-qualified} = \frac{\left|FPC_{i-Part-qualified}\right|}{\left|FPC_{i-Adaptable}\right|} \tag{4.19}$$

Using lines of code

The total number of fully qualified lines of code is divided by the total number of adaptable lines of code only. Therefore, the reusability metric is defined as:

$$RMC_{i-Part-qualified} = \frac{\left|LOCC_{i-Part-qualified}\right|}{\left|LOCC_{i-Adaptable}\right|}$$

SCENARIO 2: CBS SYSTEM-LEVEL ADAPTABLE REUSABILITY METRIC (RMC$_{CBS-PART-QUALIFIED}$)

As for fully qualified components, the partially qualified adaptable reusability metric is also calculated in three different contexts:

Case 1: Reusability metric when all parts of the component are involved.
Case 2: Reusability metric when only reused parts of the component are involved.
Case 3: Reusability metric when only adaptable parts of the component are involved.

Every case can be defined using function points as well as lines of code.

i. When all components are involved

In terms of component count

This is the case when the reusability metric is measured at system level. Here, the total number of partially qualified reused components is divided by the total number of components involved in the development. It is defined as:

$$RMC_{CBS-Part-qualified} = \frac{|C_{Part-qualified}|}{C_{CBS}} \qquad (4.20)$$

In terms of function points

In terms of function points, the reusability metric is given as the ratio of total function points and partially qualified function points of the component-based software. It is defined as:

$$RMC_{CBS-Part-qualified} = \frac{|FPC_{CBS-Part-qualified}|}{|FPC_{CBS}|} \qquad (4.21)$$

In terms of lines of code

In terms of function points, the reusability metric is given as the ratio of total lines of code and partially qualified lines of code of the component-based software. It is defined as:

$$RMC_{CBS-Part-qualified} = \frac{|LOCC_{CBS-Part-qualified}|}{|LOCC_{CBS}|}$$

ii. When only reused components are involved

In terms of component count

In this case, the ratio is taken in the context of reused components only. The total number of partially qualified adaptable components is divided by the total number of reused components involved in the development. Therefore, it is defined as:

$$RMC_{CBS-Part-qualified} = \frac{|C_{Part-qualified}|}{|C_{Reused}|} \qquad (4.22)$$

In terms of function points

The reusability metric in terms of function points is defined as the ratio between the partially qualified and total reused function points of the component-based software. That is:

$$RMC_{CBS-Part-qualified} = \frac{\left|FPC_{CBS-Part-qualified}\right|}{\left|FPC_{CBS-Reused}\right|} \tag{4.23}$$

In terms of lines of code

The reusability metric in terms of lines of code is defined as the ratio between the partially qualified and total reused lines of code of the component-based software. That is:

$$RMC_{CBS-Part-qualified} = \frac{\left|LOCC_{CBS-Part-qualified}\right|}{\left|LOCC_{CBS-Reused}\right|}$$

iii. **When only adaptable components are involved**

In terms of component count

In this case, the reusability metric is taken in the context of adaptable components only. The total number of partially qualified reused components is divided by the total number of adaptable components involved in the development and defined as:

$$RMC_{CBS-Part-qualified} = \frac{\left|C_{Part-qualified}\right|}{\left|C_{Adaptable}\right|} \tag{4.24}$$

In terms of function points

Here, the reusability metric is given as the ratio between the partially qualified and total adaptable function points of the CBS, as:

$$RMC_{CBS-Part-qualified} = \frac{\left|FPC_{CBS-Part-qualified}\right|}{\left|FPC_{CBS-Adaptable}\right|} \tag{4.25}$$

In terms of lines of code

Here, the reusability metric is given as the ratio between the partially qualified and total adaptable lines of code of the CBS:

$$RMC_{CBS-Part-qualified} = \frac{\left|LOCC_{CBS-Part-qualified}\right|}{\left|LOCC_{CBS-Adaptable}\right|}$$

4.4.5 Off-the-Shelf Reusability Metric (RM$_{Off-the-shelf}$)

Off-the-shelf components can be reused as they are, that is, without any modification. The present work considers two off-the-shelf component scenarios:

Scenario 1: *Component-level off-the-shelf reusability metric*
Scenario 2: *CBS system-level off-the-shelf reusability metric*

SCENARIO 1: COMPONENT-LEVEL OFF-THE-SHELF REUSABILITY METRIC (RMC$_{\text{i-OFF-THE-SHELF}}$)

At component level there are two different contexts for the reusability metric: when all parts of the component are involved, and when only reused parts are involved. The third context—when only adaptable components are involved—is not applicable.

Case 1: Reusability metric when all parts of the component are involved.
Case 2: Reusability metric when only reused parts of the component are involved.

Each case can be defined using function points as well as line of code.

i. When all parts of the component are involved

Using function points

In this context, we compute the reusability metric for a particular component. The total number of off-the-shelf reused function points is divided by the total number of function points of the component. It is defined as:

$$RMC_{i-\text{Off-the-shelf}} = \frac{\left|FPC_{i-\text{Off-the-shelf}}\right|}{\left|FPC_i\right|} \tag{4.26}$$

Using lines of code

The total number of off-the-shelf reused lines of code is divided by the total number of lines of code of the component. It is defined as:

$$RMC_{i-\text{Off-the-shelf}} = \frac{\left|LOCC_{i-\text{Off-the-shelf}}\right|}{\left|LOCC_i\right|}$$

ii. When only reused parts of the component are involved

Using function points

This is when the ratio is taken in the context of reused function points of the component only. Here, the total number of off-the-shelf function points is divided by the total number of reused function points. It is given as:

$$RMC_{i-\text{Off-the-shelf}} = \frac{\left|FPC_{i-\text{Off-the-shelf}}\right|}{\left|FPC_{i-\text{Reused}}\right|} \tag{4.27}$$

Using lines of code

The total number of off-the-shelf lines of code is divided by the total number of reused lines of code. It is given as:

$$RMC_{i-\text{Off-the-shelf}} = \frac{\left|LOCC_{i-\text{Off-the-shelf}}\right|}{\left|LOCC_{i-\text{Reused}}\right|}$$

SCENARIO 2: CBS SYSTEM-LEVEL OFF-THE-SHELF REUSABILITY METRIC ($RM_{CBS-OFF-THE-SHELF}$)

Off-the-shelf components are totally reusable components, without any modification. There are two cases of off-the-shelf components at system level, in terms of component count and in terms of function points.

Case 1: Reusability metric when all parts of the component are involved.
Case 2: Reusability metric when only reused parts of the component are involved.

Each case can be defined using function points as well as lines of code.

i. **When all the components are involved**
   In terms of component count
   This is the case when the ratio is taken in the context of the total number of components. In this case the total number of off-the-shelf reused components is divided by the total number of components involved in the development. It is defined as:

$$RMC_{CBS-Off-the-shelf} = \frac{|C_{Off-the-shelf}|}{|C_{CBS}|} \qquad (4.28)$$

   In terms of function points
   The reusability metric is given as the ratio between the off-the-shelf function points and the total function points of the component-based software:

$$RMC_{CBS-Off-the-shelf} = \frac{|FPC_{CBS-Off-the-shelf}|}{|FPC_{CBS}|} \qquad (4.29)$$

   In terms of lines of code
   The reusability metric is given as the ratio between the off-the-shelf lines of code and the total lines of code of the component-based software:

$$RMC_{CBS-Off-the-shelf} = \frac{|LOCC_{CBS-Off-the-shelf}|}{|LOCC_{CBS}|}$$

ii. **When only reused components are involved**
   In terms of component count
   In this case, the ratio is taken in the context of reused components only. In this case the total number of off-the-shelf reused components is divided by the total number of reused components involved in the development. It is defined as:

$$RMC_{Off-the-shelf} = \frac{|C_{Off-the-shelf}|}{|C_{Reused}|} \qquad (4.30)$$

   In terms of function points
   In terms of function points, the reusability metric is achieved by dividing total off-the-shelf function points by total reused function points of the component-based software. It is defined as:

$$RMC_{CBS-Off-the-shelf} = \frac{|FPC_{CBS-Off-the-shelf}|}{|FPC_{CBS-Reused}|} \qquad (4.31)$$

In terms of lines of code

In terms of lines of code, the reusability metric is achieved by dividing total off-the-shelf lines of code by total reused lines of code of the component-based software. It is defined as:

$$RMC_{CBS-Off-the-shelf} = \frac{|LOCC_{CBS-Off-the-shelf}|}{|LOCC_{CBS-Reused}|}$$

4.5 Component Selection and Verification Using Reusability Metrics

Using the proposed reusability metrics, we can select and verify the components in component-based software. If we have to single out a component from a number of available components, there should be some selection and verification criteria. In this section, we propose a selection procedure for a component and verification of that selection in component-based software. We can store values of the reusability metric in the performance coefficient list of the component. Using coefficients of performance, we form a matrix for each component called the reusability matrix.

4.5.1 Reusability Matrix

The reusability matrix is a matrix containing the reusability metric computations of each component in the component-based software. Information containing the reusability matrix works as a selection and verification criterion for the candidate components.

The reusability matrix is a row-column matrix containing the values of size estimations of components in the form of function points or lines of code. The columns represent the candidate component's name and the rows represent the function points of the corresponding components, as shown in Table 4.1.

TABLE 4.1

Reusability Matrix

Reusability matrix	Candidate Components			
	C_1	C_2	C_3	C_4
$RMC_{i-Part-qualified}$	Partially qualified RM value of component C_1	Partially qualified RM value of component C_2	Partially qualified RM value of component C_3	Partially qualified RM value of component C_4
$RMC_{i-Full-qualified}$	Fully qualified RM value of component C_1	Fully qualified RM value of component C_2	Fully qualified RM value of component C_3	Fully qualified RM value of component C_4
$RMC_{i-Off-the-shelf}$	Off-the-shelf RM value of component C_1	Off-the-shelf RM value of component C_2	Off-the-shelf RM value of component C_3	Off-the-shelf RM value of component C_4

4.5.2 Case Study

This section discusses a case study analyzing the reusability metric with the help of the reusability matrix. The study concerns a part of some component-based software.

Figure 4.1 shows the component diagram of the Login_Process and Exit_Process of a software that has five components: UserLoginComponent, LoginVerificationComponent, UserLoginDatabase, UserPageComponent, and ExitComponent. The user fills data/information related to their login in the UserLoginComponent, which is checked and cleared by the LoginVerificationComponent. Verification and clearance is done through the UserLoginDatabase component. If the user data is verified then the UserPageComponent is enabled, otherwise LoginVerificationComponent redirects the UserLoginComponent. Here the aim is to select a suitable UserPageComponent (shown in oval) component.

This UserPageComponent has five candidate components, denoted as C_{11}, C_{12}, C_{13}, C_{14} and C_{15}, which we have to test for reusability, as shown in Table 4.2. Here the assumption is that function points are available with each candidate component. Total function points of candidate component C_{11} are 600, C_{12} has 620 function points, C_{13} has 580, C_{14} has 570 and C_{15} has 480 function points, as shown in Table 4.2. From these function points, the developer can identify the partially qualified, fully qualified and off-the-shelf function points of the component, as required in the software under development. In this case study,

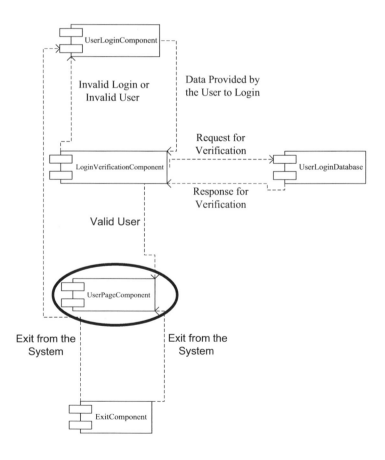

FIGURE 4.1
User login page.

TABLE 4.2

Partially Qualified, Fully Qualified and Off-the-Shelf Function Points of Five Candidate Components

Function Points	C_{11}	C_{12}	C_{13}	C_{14}	C_{15}
$|FPC_i|$	600	620	580	570	480
$|FPC_{i\text{-Part-qual}}|$	140	220	100	300	140
$|FPC_{i\text{-Full-qual}}|$	300	230	280	120	200
$|FPC_{i\text{-Off-the-shelf}}|$	100	90	130	60	50

TABLE 4.3

Reused, Adaptable and New Function Points of Five Candidate Components

Function Points	C_{11}	C_{12}	C_{13}	C_{14}	C_{15}
$|FPC_{i\text{-Reused}}|$	540	540	510	480	390
$|FPC_{i\text{-Adaptable}}|$	440	450	380	420	340
$|FPC_{i\text{-New}}|$	60	80	70	90	90

the values of partially qualified, fully qualified and off-the-shelf function points of the component are taken, as shown in Table 4.2. Here, the columns represent the candidate components' names, C_{11}, C_{12}, C_{13}, C_{14} and C_{15}, and the rows represent the function points of the corresponding components.

Using Table 4.2, and Equations (4.1), (4.2), (4.3) and (4.4) respectively, we can compute the values of the number of reused, adaptable and new function points of the candidate components, as shown in Table 4.3.

With the help of Tables 4.2 and 4.3, now we can draw the reusability matrix for five candidate components using the reusability-metric method.

CASE 1   REUSABILITY MATRIX WHEN ALL PARTS OF THE COMPONENTS ARE INVOLVED IN THE COMPUTATION

Applying the values given in Tables 4.2 and 4.3 to Equations (4.5), (4.17), (4.8), and (4.26), respectively, we assess the values of the reusability metrics of component C_i, in the context when calculations are made in terms of all parts of the component, that is, new and reused. These computations are shown in Table 4.4.

TABLE 4.4

Reusability Matrix when New and Reused Function Points are Involved

Reusability Matrix	C_{11}	C_{12}	C_{13}	C_{14}	C_{15}
RMC_i	**0.9**	0.87	0.88	0.84	0.81
$RMC_{i\text{-Part-qual}}$	0.23	0.35	0.17	**0.53**	0.29
$RMC_{i\text{-Full-qual}}$	0.5	0.37	**0.48**	0.21	0.42
$RMC_{i\text{-Off-the-shelf}}$	0.17	0.14	**0.22**	0.10	0.10

TABLE 4.5

Reusability Matrix when only Reused Function Points of the Component are Involved

Reusability Matrix	C_{11}	C_{12}	C_{13}	C_{14}	C_{15}
$RMC_{i\text{-part-qualified}}$	0.26	0.41	0.20	**0.62**	0.36
$RMC_{i\text{-full-qualified}}$	**0.55**	0.42	**0.55**	0.25	0.51
$RMC_{i\text{-off-the-shelf}}$	0.18	0.17	**0.25**	0.12	0.13

CASE 2:   REUSABILITY MATRIX WHEN ONLY REUSED FUNCTION POINTS ARE INVOLVED

Applying the values given in Tables 4.2 and 4.3 to Equations (4.18), (4.9), and (4.27), respectively, we calculate the values of reusability metrics of component C_i, in the context of reused function points only, as shown in Table 4.5.

CASE 3:   REUSABILITY MATRIX WHEN ONLY ADAPTABLE FUNCTION POINTS ARE INVOLVED

Applying the values given in Tables 4.2 and 4.3 to Equations (4.19) and (4.10) respectively, we calculate the values of reusability metrics of component C_i, in the context of adaptable function points only.

4.5.3 Selection and Verification of Components

After calculating the reusability-matrix values, we now define the candidate component selection procedure. From the reusability matrices shown in Tables 4.4, 4.5, and 4.6, we note that each row contains the value of function points of the corresponding candidate components. For example, the first row of Table 4.4 contains the reusability-metric values of component C_i (RMC_i) of the corresponding candidate components from C_{11} to C_{15}. We also note that component C_{11} has the highest reusability-metric value in the context of the whole component. We select component C_{11} when reusability is considered in the context of all parts of the component. From the reusability matrices computed in Tables 4.4, 4.5, and 4.6, we can select and verify the selection of components.

TABLE 4.6

Reusability Matrix when only Adaptable Function Points of the Component are Involved

Reusability matrix	C_{11}	C_{12}	C_{13}	C_{14}	C_{15}
$RMC_{i\text{-part-qualified}}$	0.32	**0.49**	0.26	0.71	0.41
$RMC_{i\text{-full-qualified}}$	0.68	0.51	**0.74**	0.29	0.59

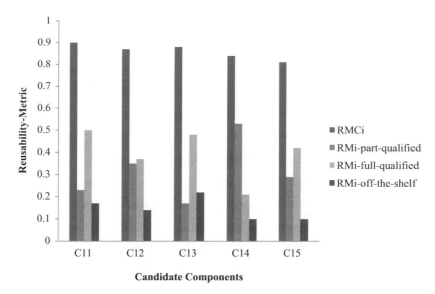

FIGURE 4.2
Reusability graph when components contain new and reused function points.

4.5.3.1 Selection of Components When All the Parts of the Component Are Considered

From Table 4.4 and Figure 4.2 we note that, when selection is made at component level, then the eligible components are:

a. Highest reusability-metric value: *Component C_{11}*
b. Highest partially qualified component: *Component C_{14}*
c. Highest fully qualified component: *Component C_{13}*
d. Highest off-the-shelf component: *Component C_{13}*

4.5.3.2 Selection of Components When Reused Components Are Considered

From Table 4.5 and Figure 4.3, we note that when selection is made at reused component level, then the eligible components are:

a. Highest partially qualified component: *Component C_{14}*
b. Highest fully qualified component: *Component C_{11} and Component C_{13}*
c. Highest off-the-shelf component: *Component C_{13}*

4.5.3.3 Selection of Components When Adaptable Components Are Considered

From Table 4.6 and Figure 4.4, we note that when selection is made at adaptable component level, then the eligible components are:

a. Highest partially qualified component: *Component C_{12}*
b. Highest fully qualified component: *Component C_{11} and Component C_{13}*

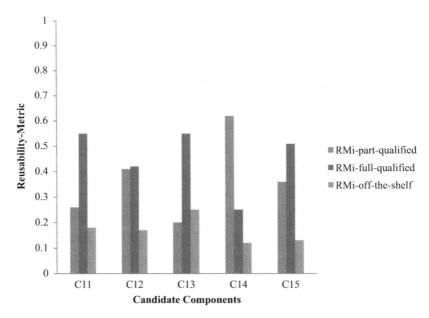

FIGURE 4.3
Reusability graph when components contain reused function points only.

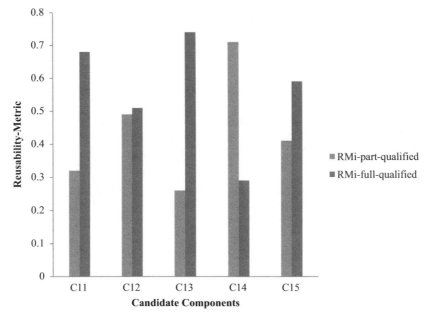

FIGURE 4.4
Reusability graph when components contain adaptable function points only.

We can select components according to our criteria and selection level. The selection process for components at system level can be defined in a similar manner. At system level we store all the values in terms of applications, and all the computations are in terms of function points defined at CBS-system level.

Summary

In this chapter, estimation methods for reusability of the components in the component-based software environment are elaborated for various categories of reusable components. The four classes of components are *new*, *partially qualified*, *fully qualified* and *off-the-shelf* components. The reusability metric for these identified components at two different levels—system level and individual component level—are defined. Further, this work defines the reusability matrix for a number of candidate components serving the same purpose. Using above defined reusability metric and reusability matrix, one can verify the selection of an individual component from the candidate components.

A case study has been used to model a scenario of five components. These components are participating for the same purpose and we have to select one among them for reuse in software that is under development. The results obtained from this case study illustrate that the reusability metric can be used at three different levels for component selection as well as verification. The values of different reusability matrices can be used to select or reject the candidate component. The results suggest that an increase in the reusability of a component results in enhanced probability of selection of that component. These metrics and matrices are helpful not only for computations of the reusability attributes of components in CBS applications but can also be stored as a performance coefficient along with the component in the repository.

References

Albrecht, A. and J. E. Gaffney. 1983. "Software Function Source Line of code and Development Effort Prediction: A Software Science Validation." *IEEE Transactions on Software Engineering*, SE-9: 639–648.

Atkinson, C. et al. 2002. *Component-Based Product-Line Engineering with UML*. Addison-Wesley, London.

Basili, V.R. and H. D. Rombach. 1988. "Towards a Comprehensive Framework for Reuse: A Reuse-Enabling Software Evolution Environment." *Technical Report CS-TR-2158*, University of Maryland.

Bersoff E.H. and A. M. Davis. 1991. "Impacts of Life Cycle Models on Software Configuration Management." *Communications of the ACM*, 8(34): 104–118.

Biggerstaff, T. J. and C. Richter. 1989. *Reusability Framework, Assessment, and Directions, Software Reusability: Concepts and Models*. Addison-Wesley Publishing Company, New York, 1–18.

Braun, C. L. 1994. "Reuse." In *Marciniak*, IEEE, 1055–IEEE, 1069.

Brown, A. W. and K. C. Wallnau. 1998. "The Current State of CBSE." *IEEE Software*, 15(5): 37–46.

Cooper, J. 1994. "Reuse—The Business Implications." In *Marciniak*, 1071–1077.

Freeman, P. 1987. "Reusable Software Engineering Concepts and Research Directions." In *Tutorial: Software Reusability*, ed. P Freeman. IEEE Computer Society Press, Los Alamitos, 10–23.

Johnson, L. and D. R. Harri. 1991. Sharing and Reuse of Requirements Knowledge," In *Proceedings of the KBSE-91*. IEEE Press, Los Alamitos, CA, 1991, pp. 57–66.

Kim, Y. and E. A. Stohr. 1998. "Software Reuse: Survey and Research Directions." *Journal of Management Information Systems*, 14: 113–147.

Krueger, C. W. 1992. "Software Reuse." *ACM Computing Surveys (CSUR)*, 24(2): 131–183.

Lim, W. C. 1994. "Effects of Reuse on Quality, Productivity, and Economics." *IEEE Software*, 11(5): 23.

Maiden, N. and A. Sutcliff. 1991. Analogical Matching for Specification Reuse." In *Proceedings of KBSE-91*, IEEE Press, Los Alamitos, CA, 108–116.

Matsumoto, Y. 1989. *Some Experiences in Promoting Reusable Software: Presentation in Higher Abstract Levels, Software Reusability: Concepts and Models*. ACM Addison-Wesley Publishing Company, New York, 157–185.

Maxim, R. B. 2019. *"Software Reuse and Component-Based Software Engineering."* CIS 376, UM-Dearborn.

McClure, C. 1989. *CASE Is Software Automation*. Prentice Hall, Englewood Cliffs, NJ.

McIlroy, M. D. 1976. "Mass Produced Software Components. In J.M. Buxton, P. Naur, and B. Randell, eds., *Software Engineering Concepts and Techniques*." *NATO Conference on Software Engineering*, NATO Science Committee, Garmisch, Germany 88–98.

Prieto-Diaz, R. 1992. *Classification of Reusable Modules, Software Reusability: Concepts and Models*, J. B. Ted and J. P. Alan, eds. Addison-Wesley Publication Company, New York, 99–123.

Tracz, W. 1995. *Confessions of a Used Program Salesman: Institutionalizing Software Reuse*. Addison-Wesley, Reading, MA.

5

Interaction and Integration Complexity Metrics for Component-Based Software

5.1 Introduction

In component-based software development, the developer's emphasis is on the assembly and integration of pre-constructed, pre-examined, customizable and easily deployable software components, rather than on developing software from scratch. Complexity assessment is always an exigent task for the designers of large-scale applications and software where software is divided into various components or modules. Complexity accumulates as the size of the software increases. Integration and interactions among the components follow a well-defined architectural design and should occur according to the user's requirement specification. Components interact with each other for two basic reasons:

1. To access the services and functionalities of other components, and
2. To provide services and their own functionalities to other components.

These interactions among various components generate complexity in component-based software. In general, the term complexity is defined as the assessment and calculation (sometimes prediction) of resources required to resolve a problem. In the domain of software development, complexity is an attribute which cannot be measured directly, that is, complexity is an indirect measure. In general terms, complexity is the assessment of hardware and software resources needed by software. In software development, complexity is treated as an indirect measurement, unlike direct measurements like lines of code or cost estimation (Krueger 1992). Internal as well as external interactions play a major role in software complexity.

In the context of software development, interaction behavior of various parts of a program is used to measure complexity. These parts may be single-line code, a group of lines of code (functions), a group of functions (modules) or ultimately components. As the size of parts of the software increases, the total number of interactions will also increase, as well as the complexity. Software applications are composed of dependent or independently deployable components. These components are assembled for the purpose of contributing their functionalities to the system. Technically this assembly is referred to as integration of and interaction among components. We have a sufficient number of measures and metrics to assess the complexity of standalone programs as well as small-sized conventional software, suggested and practiced by numerous practitioners (Bauer et al. 1968, Boehm 1981,

IEEE 1987, Schach 1990, IEEE 1990, Gaedke and Rehse 2000, Tiwari and Kumar 2014). In the literature, complexity of programs and software is treated as a "multidimensional construct" (Boehm 1981, Shepperd 1988).

Interaction and integration complexities of various pieces of code play a vital role in the overall behavior of software. As the code count increases, the interaction level of the software also increases in accordance with its requirements.

This chapter aims to suggest and develop competent and proficient interaction complexity estimation measurements and metrics. Two complexity computation techniques are suggested:

1. In-out interaction complexity
2. Cyclomatic complexity for component-based software

5.2 In-Out Interaction Complexity

Methods and metrics proposed so far in the literature are defined on the basis of interactions among instructions, operations, procedures and functions of individual and stand-alone programs and codes. These metrics are appropriate for small-sized codes. Some measures are also defined for object-oriented software, but they are not adequate for component-based applications. In component-based software engineering, components have connections and communications with each other to exchange services and functionalities. Interaction edges are used to denote the connections among components. In this section we define some simple metrics to assess the interaction of component-based software. Metrics defined in this work consider the individual interactions of components as well as inter-component interactions. These metrics are helpful for exploring the non-functional attributes of components and component-based software.

When components in the component-based software interact with each other, they generate interaction complexities. These complexities are vital not only for designers and testers but also for maintenance activities. As the interaction complexity increases, related factors like usability, reusability, understandability and similar factors may be affected.

5.2.1 In-Out Interaction Complexity Graph

This chapter describes in-out interaction complexity metrics to assess and evaluate the complexity produced by interactions among different components. Complexities generated due to in-out interactions are easy to compute and helpful for designers and developers in terms of keeping track of the complexity level of components and component-based software.

5.2.2 Notations

Graph-theory notation is used throughout this chapter to denote all types of interactions shared among various components. Components are shown as vertices, and interactions among components are denoted as edges. Generally, components make interactions to share and communicate information, data, control or similar resources.

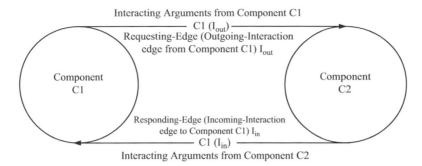

FIGURE 5.1
Interactions between two components.

For in-out interaction metrics, edges are divided into two categories:

- Incoming interactions
- Outgoing interactions

  i. **Incoming interactions:** Edges in the graph that are coming towards the component are referred to as incoming interactions. These edges are drawn due to:

  - A request sent by any other component to the receiving component, or
  - A response from some other component to the request-sending component

  Figure 5.1 shows incoming interactions and outgoing interactions. The edge coming from component C2 in the direction of component C1 is termed the incoming interaction for component C1 and denoted as I_{in}.

  ii. **Outgoing interactions:** The edges in the graph that are going out from the component are referred to as the outgoing interactions. These edges are drawn due to either:

  - A request sent by the component to some other component, or
  - A response from the component to the request-sending component.

  In Figure 5.1, the edge going from component C1 in the direction of component C2 is termed the outgoing interaction for component C1 and denoted as I_{out}.

When we assess in-out interaction complexity, all types of edges are considered, both incoming and outgoing interactions. From Figure 5.1 we can observe that component C1 contains one outgoing interaction (I_{out}) as a requesting-edge from C1 to C2 and one incoming interaction (I_{in}) as a response-edge coming from C2 to C1. As components C1 and C2 are both inter-dependent for requests and responses, evaluation of the in-out interaction complexity mustr also consider complexities produced by both components C1 and C2.

5.2.3 Total Interactions of a Component

For a particular component "C," total interactions can be defined as the summation of the total number of incoming interactions (I_{in}) and total number of outgoing interactions (I_{out}):

$$TI_{Ci} = I_{out} + I_{in} \tag{5.1}$$

where I_{in} is the incoming interaction and I_{out} is the outgoing interaction of component "C."

5.2.4 Total Interactions of Component-Based Software

Component-based software consists of a number of components interacting with each other through various incoming and outgoing interactions. the total interactions of component-based software is defined as the summation of the total number of incoming interactions (I_{in}) and total number of outgoing interactions (I_{out}) of all the contributing components:

$$TI_{CBS} = \sum_{i=1}^{n} I_{out} + \sum_{i=1}^{n} I_{in} \tag{5.2}$$

where n is the total count of contributing components in the component-based software.

5.2.5 Interaction Ratio of a Component

The interaction-ratio metric represents the amount of dependency of individual components that are making interactions in the software. The interaction ratio of a component can be defined as the ratio between the total outgoing interactions (I_{out}) and the total incoming interactions (I_{in}). It is defined as:

$$IR_{Ci} = \frac{I_{out}}{I_{in}} = \begin{cases} <1 & I_{out} < I_{in}, \text{high dependency.} \\ =1 & I_{in} = I_{out}, \text{equal dependency.} \\ >1 & I_{out} > I_{in}, \text{high dependency.} \end{cases} \tag{5.3}$$

We can verify that

- If $IR_{Ci} = 1$ when $I_{in} = I_{out}$, the total number of incoming interactions is equal to the total number of outgoing interactions. We can conclude that there is equivalent dependency among contributing components.
- If $IR_{Ci} < 1$ it means that $I_{out} < I_{in}$, that is, the total number of incoming interactions is greater than the total number of outgoing interactions. We can infer the result that there is high dependency on the responding component.
- If $IR_{Ci} > 1$ it means $I_{out} > I_{in}$, implying that there is high dependency on the responding component.

5.2.6 Interaction Ratio of Component-Based Software

The interaction-ratio metric of component-based software (IR_{CBS}) denotes the amount of inter-dependency among contributing components that are making interactions. The interaction ratio of component-based software can be defined as the ratio between the total number of outgoing interactions (I_{out}) and the total number of incoming interactions (I_{in}) made by all the contributing components of the software. It is given as:

$$IR_{CBS} = \frac{\sum_{i=1}^{n} I_{out}}{\sum_{i=1}^{n} I_{in}} \tag{5.4}$$

where n is the total of all contributing components in the component-based software.

5.2.7 Average Interactions Among Components

The average-interaction metric represents the amount of connectivity among participating components. The average interaction among contributing components is defined as the ratio of the summation of incoming interactions (I_{in}) and outgoing interactions (I_{out}) to the total contributing components in the component-based software. The total of contributing components is denoted C_n. Average interactions is defined as:

$$\text{AI}_{C_n} = \frac{(I_{in} + I_{out})}{C_n} = \begin{cases} < \dfrac{1}{2}, \text{half of the components are disjoint.} \\ = 1, \text{ atleast one interaction among components.} \\ > 1, \text{ coupling is very high among components.} \end{cases} \quad (5.5)$$

We can verify that,

- If $\text{AI}_{C_n} < \tfrac{1}{2}$, it implies that at least half of the interacting components are disjoint,
- If $\text{AI}_{C_n} = 1$, it denotes that there is at least one interaction among components, and
- If $\text{AI}_{C_n} > 1$, then it is concluded that components are highly coupled.

5.2.8 Interaction Percentage of a Component

The interaction-percentage metric is used to represent the degree of underflow or overflow of interactions amongst contributing components. The interaction percentage of a component IP_{C_n} is the ratio of the summation of in interactions (I_{in}) and out interactions (I_{out}) made by the particular component, to the total interactions made by all the contributing components. Total interactions of components are defined in Section 5.2.1 and total interactions made by all the contributing components is defined in Section 5.2.2. The interaction percentage of a component is defined as:

$$\text{IP}_{C_n} = \frac{(I_{in} + I_{out})}{\text{TI}_{CBS}} = \begin{cases} < 1, \text{ underflow condition.} \\ = 1, \text{ balanced condition.} \\ > 1, \text{ overflow condition.} \end{cases} \quad (5.6)$$

We can verify that

- If $IP_{C_n} < 1$, it implies an underflow situation, that is, more interactions are possible among the components,
- If $IP_{C_n} = 1$, it shows a balanced situation, and
- If $IP_{C_n} > 1>$, it shows an overflow situation, that is, components are sharing heavy interaction, which will increase the complexity.

5.2.9 Case Study

To implement the in-out interaction complexity, we consider two exemplar case studies, the first having two components, and the second having four components.

CASE 1: IN-OUT INTERACTION COMPLEXITY GRAPH WITH TWO COMPONENTS

This is the simplest case, when the in-out interaction graph consists of only two components, C1 and C2, as shown in Figure 5.1.

Total Interactions of a Component Ci (TI_{Ci}):

Applying the values of Table 5.1 in Equation (5.1), we get total interactions of component C1 and component C2 as,

$$TI_{C1} = I_{out} + I_{in} = 1 + 1 = 2$$

$$TI_{C2} = I_{out} + I_{in} = 0 + 0 = 0$$

Total Interactions of Component-Based Software (TI_{CBS}):

This software consists of two components only, C1 and C2. Applying the values given in Table 5.1 in Equation (5.2), we get,

$$TI_{CBS} = \sum_{i=1}^{n} I_{out} + \sum_{i=1}^{n} I_{in} = (1+0) + (1+0) = 2$$

Interaction Ratio of a Component "Ci" (IR_{Ci}):

Applying the values of Table 5.1 in Equation (5.3), we get the interactions ratio of component C1 and component C2.
    For component C1:

$$IR_{Ci} = \frac{I_{out}}{I_{in}},$$

$$IR_{C1} = 1/1 = 1$$

For component C2, there are no incoming-outgoing interactions. Hence the interaction ratio is not applicable to component C2.

Interaction Ratio of Component-Based Software (IR_{CBS}):

Case 1 consists of two components, C1 and C2. Applying the values given in Table 5.1 in Equation (5.4), we get,

$$IR_{CBS} = \frac{\sum_{i=1}^{n} I_{out}}{\sum_{i=1}^{n} I_{in}} = 1/1 = 1$$

TABLE 5.1

Number of In-Out Interactions between Components C1 and C2

Component	Incoming Interactions (I_{in})	Outgoing Interactions (I_{out})
C1	1	1
C2	0	0

Average Interactions Among Components (AI_{C_n})

In case 1, the number of components is two, hence to find the average interactions, we apply Equation (5.5) as,

$$AI_{C_n} = \frac{\left(I_{in} + I_{out}\right)}{C_n} = 2/2 = 1$$

Interaction Percentage of Component C (IP_{C_n}):

We apply the values defined in Table 5.1 to Equation (5.6) to compute the interaction percentage.

Interaction Percentage of Component C1:

$$IP_{C_1} = \frac{\left(I_{in} + I_{out}\right)}{TI_{CBS}} = 2/2 = 1$$

Interaction Percentage of Component C2:

$$IP_{C_2} = \frac{\left(I_{in} + I_{out}\right)}{TI_{CBS}} = 0/12 = 0$$

CASE 2: IN-OUT INTERACTION COMPLEXITY GRAPH WITH FOUR COMPONENTS

In this case, four components are used to represent interactions among them. Every participating component has associated in-out interactions as shown in Figure 5.2.

In the given in-out complexity graph, component C1 is interacting with components C2 and C4. Component C2 is not initiating any interaction, rather it is replying to the request made by component C4. Component C3 has been integrated with C1 and C4. Component C4 is integrating with C3 and C1. All incoming interactions and outgoing interactions are shown in the figure.

Table 5.2 shows the values of in-out interactions of components participating in the software.

Total Interactions of a Component Ci (TI_{Ci}):

Applying the values of Table 5.2 in Equation (5.1), we get total interactions of the participating component as

$$TI_{C1} = I_{out} + I_{in} = 2 + 2 = 4$$

$$TI_{C2} = I_{out} + I_{in} = 0 + 0 = 0$$

$$TI_{C3} = I_{out} + I_{in} = 2 + 2 = 4$$

$$TI_{C4} = I_{out} + I_{in} = 2 + 2 = 4$$

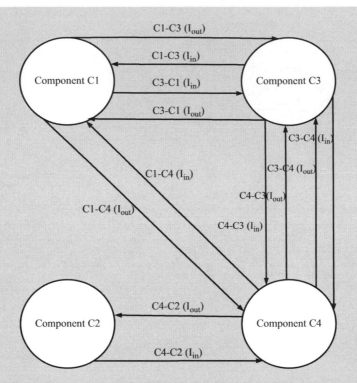

FIGURE 5.2
Interactions among four components.

TABLE 5.2

Number of In-Out Interactions among Four Components

Component	Incoming Interactions (I_{in})	Outgoing Interactions (I_{out})
C1	2	2
C2	0	0
C3	2	2
C4	2	2

Total Interactions of Component-Based Software (TI_{CBS}):

This software consists of four components and they are interacting with each. Applying the values given in Table 5.2 in Equation (5.2), we get,

$$TI_{CBS} = \sum_{i=1}^{n} I_{out} + \sum_{i=1}^{n} I_{in}$$
$$= (2+0+2+2) + (2+0+2+2) = 6+6 = 12$$

Interaction Ratio of a Component "Ci" (IR_{Ci}):

Applying the values of Table 5.2 in Equation (5.3), we get the interactions ratio of components C1, C2, C3 and C4.

For component C3:

$$IR_{Ci} = \frac{I_{out}}{I_{in}},$$

$$IR_{C1} = 2/2 = 1$$

For component C2, there are no incoming–outgoing interactions. Hence the interaction ratio is not applicable to component 2.

For component C3:

$$IR_{Ci} = \frac{I_{out}}{I_{in}},$$

$$IR_{C3} = 2/2 = 1$$

For component C4:

$$IR_{Ci} = \frac{I_{out}}{I_{in}},$$

$$IR_{C4} = 2/2 = 1$$

Interaction Ratio of Component-Based Software (IR_{CBS}):

We can note that case 2 consists of four components; there is a total of six incoming interactions in the software , and a total of six outgoing interactions. Applying values given in Table 5.2 in Equation (5.4), we now get

$$IR_{CBS} = \frac{\sum_{i=1}^{n} I_{out}}{\sum_{i=1}^{n} I_{in}} = 6/6 = 1$$

Average Interactions Among Components (AI_{Cn})

In Case 1, the total number of contributing components is four, hence to find average interactions, we apply Equation (5.5):

$$AI_{C_n} = \frac{(I_{in} + I_{out})}{C_n} = (6+6)/4 = 3$$

Interaction Percentage of Component C (IP_{Cn}):

We apply the values defined in Table 5.2 to Equation (5.6) to compute the interaction percentage.

Interaction Percentage of Component C1:

$$IP_{C_1} = \frac{(I_{in} + I_{out})}{TI_{CBS}} = 4/12 = 0.33$$

Interaction Percentage of Component C2:

$$IP_{C_2} = \frac{(I_{in} + I_{out})}{TI_{CBS}} = 0/12 = 0$$

Interaction Percentage of Component C3:

$$IP_{C_3} = \frac{(I_{in} + I_{out})}{TI_{CBS}} = 4/12 = 0.33$$

Interaction Percentage of Component C4:

$$IP_{C_4} = \frac{(I_{in} + I_{out})}{TI_{CBS}} = 4/12 = 0.33$$

5.3 Cyclomatic Complexity for Component-Based Software

Thomas J. McCabe (1976) defined a complexity measurement method based on the interactions among the statements of a program. He offered and implemented graph-theoretic notions in programming applications. He used a code control-flow graph to compute the complexity. In a control-flow graph, a sequential block of code or a single statement is represented as a node, and control flows among these nodes are represented as edges. A cyclomatic complexity metric is easy to compute and maintain, as well as giving relative complexity of various designs. This method is applicable to standalone programs and to hierarchical nests of subprograms.

McCabe used graph-theoretic notations to draw a control-flow graph where a graph denoted G has n number of nodes, e number of connecting edges and p number of components.

The cyclomatic complexity, V(G), is calculated as

$$V(G) = e - n + 2p,$$

where 2 is the "result of adding an extra edge from the exit node to the entry node of each component module graph" (Pressman 2005). In a structured program where we have predicate nodes, complexity is defined as

$$V(G) = \text{number of predicate nodes} + 1,$$

where predicate nodes are the nodes having two and only two outgoing edges.

In his implementations, McCabe defined a program cyclomatic complexity value of less than 10 as reasonable. If a program has a hierarchical structure, that is, one subprogram is calling another, the cyclomatic complexity is the summation of individual complexities of these two subprograms and is given as

$$V(G) = v(P_1 + P_2) = v(P_1) + v(P_2),$$

where P_1 and P_2 are two subprograms and P_1 is calling P_2.

Complexity depends not on the size, but on the coding structure of the program. If a program has only one statement then it has complexity 1. That is, $V(G) \geq 1$. Cyclomatic complexity $V(G)$ actually defines the number of independent logics/paths in the program.

5.3.1 Flow Graph for Component Integration

Component-based software applications are composed of independently deployable components. The assembly of these components has a common intention to contribute their functionalities to the system. Technically this assembly is referred to as integration of and interaction among components. Interaction edges are used to denote the connections among components, with an edge for each requesting communication and similarly an edge for each responding communication. A requesting component sends a "request edge" and the responding component sends a "response edge." This metric is based on the concept of double edges to show the interaction among requesting and responding components. In a flow graph, vertices denote components, and edges between components are used to denote integrations and interactions among them. The internal structures of components are also shown in the control-flow graph. Figure 5.3 shows a flow graph where two components are integrated.

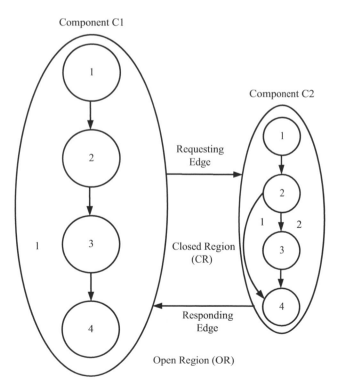

FIGURE 5.3
Interaction flow graph of two components.

5.3.2 Cyclomatic Complexity Metric

Component-based software is complex and includes a large number of independent and context-free components. Their behavior and nature are different from those of small-scale applications and software. We cannot simply apply a cyclomatic complexity formula that was developed for programs, to component-based software. This section sets out two methods of computing the cyclomatic complexity of component-based software.

- **Method 1:**
  For such multifaceted and multi-component software, cyclomatic complexity is defined as:

$$V(G) = |E| - |V| + 2 + |P| \tag{5.7}$$

  where

  $|E|$ denotes number of edges, $|V|$ is used to denote total number of vertices,

  $|P|$ is the total of contributing components and

  Constant 2 is used to indicate that "the node V contributes to the complexity if its out-degree is 2."

- **Method 2:**
  Another metric for computing the cyclomatic complexity of component-based software is defined in Equation (5.8) as:

$$V(G) = \sum_{i=1}^{n} (IC)_i + \sum_{j=1}^{m} (CR)_j + OR \tag{5.8}$$

  where

  $(IC)_i = (IC_1, IC_2, IC_3, \ldots, IC_n)$ denoting cyclomatic complexities of all the contributing components,

  $(CR)_j = (CR_1, CR_2, CR_3, \ldots, CR_m)$ denoting the total closed regions found in the graph, and

  OR denotes the open region that is always 1.

5.3.3 Case Study

To illustrate these metrics, example case studies with different scenarios are discussed. Both the methods defined in Equations (5.7) and (5.8) are applied to all scenarios.

SCENARIO 1:   INTERACTION SCENARIO BETWEEN TWO COMPONENTS

Figure 5.3 shows an interaction scenario between two components. The components' internal structures are also shown.

Method 1:
From Figure 5.6, it is noted that

$$|E| = 10, \ |V| = 9, \text{ and } |P| = 2, \text{ therefore,}$$

$$V(G) = |E| - |V| + 2 + |P|$$

$$= 10 - 9 + 2 + 2 = 5$$

Method 2:
From Figure 5.3, it is noted that,

$$n = 2, \ m = 1, \ (IC)_1 = 1, \ (IC)_2 = 2, \ CR = 1, \ OR = 1, \text{ therefore,}$$

$$V(G) = \sum_{i=1}^{n} (IC)_i + \sum_{j=1}^{m} (CR)_j + OR$$

$$V(G) = (1 + 2) + 1 + 1 = 5$$

SCENARIO 2: INTERACTION SCENARIO BETWEEN THREE COMPONENTS

Figure 5.4 shows an interaction scenario between three components. The internal structures of components are also shown in the figure.

Method 1:
From Figure 5.4, it is noted that,
$|E| = 22$, $|V| = 16$, and $|P| = 3$, therefore,

$$V(G) = |E| - |V| + 2 + |P|$$

$$= 22 - 16 + 2 + 3 = 11$$

Method 2:
From Figure 5.4, it is noted that,

$$n = 3, \ m = 2, \ (IC)_1 = 2, \ (IC)_2 = 2, \ (IC)_3 = 2, \ CR = 1, \ OR = 1, \text{ therefore,}$$

$$V(G) = \sum_{i=1}^{n} (IC)_i + \sum_{j=1}^{m} (CR)_j + OR$$

$$V(G) = (2 + 2 + 4) + 2 + 1 = 11$$

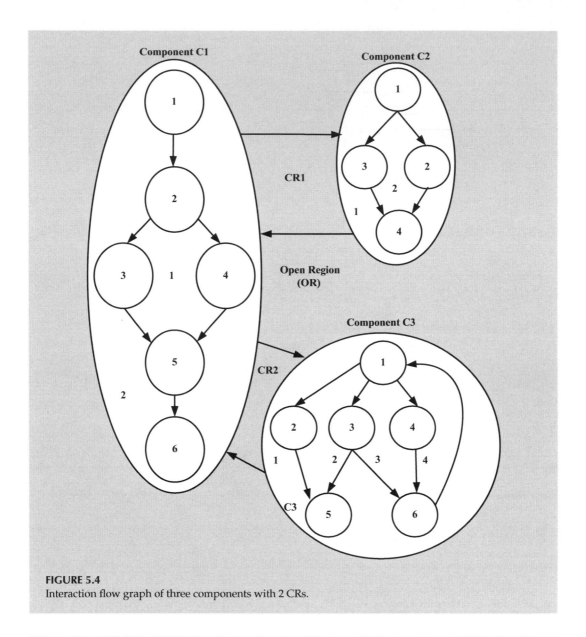

FIGURE 5.4
Interaction flow graph of three components with 2 CRs.

SCENARIO 3: INTERACTION SCENARIO BETWEEN THREE COMPONENTS AND 4 CRs

Figure 5.5 shows an interaction scenario between three components. The internal structures of components are also shown in the figure.

Method 1:
From the Figure 5.5, it is noted that,

$$|E| = 24, \ |V| = 16, \ \text{and} \ |P| = 3, \ \text{therefore,}$$

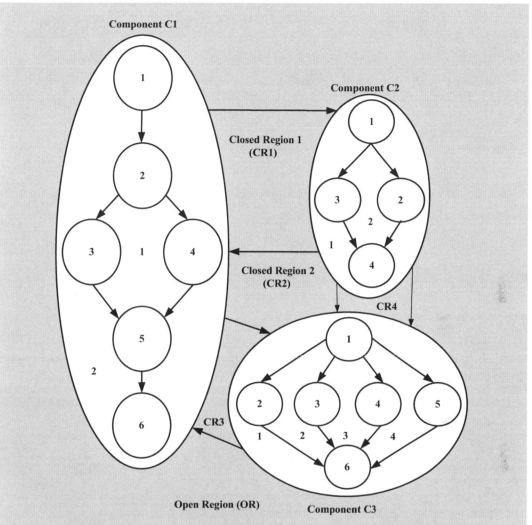

FIGURE 5.5
Interaction flow graph of three components with 4 CRs.

$$V(G) = |E| - |V| + 2 + |P|$$

$$= 24 - 16 + 2 + 3 = 13$$

Method 2:
From the Figure 5.5, it is noted that,

$$n = 3, \ m = 4, \ (IC)_1 = 2, \ (IC)_2 = 2, \ (IC)_3 = 4, \ CR = 1, \ OR = 1, \ \text{therefore,}$$

$$V(G) = \sum_{i=1}^{n} (IC)_i + \sum_{j=1}^{m} (CR)_j + OR$$

$$V(G) = (2 + 2 + 4) + 4 + 1 = 13$$

SCENARIO 4:   INTERACTION SCENARIO AMONG FOUR COMPONENTS AND 3 CRs

Figure 5.6 shows an interaction scenario between four components. The internal structures of components are also shown in the figure.

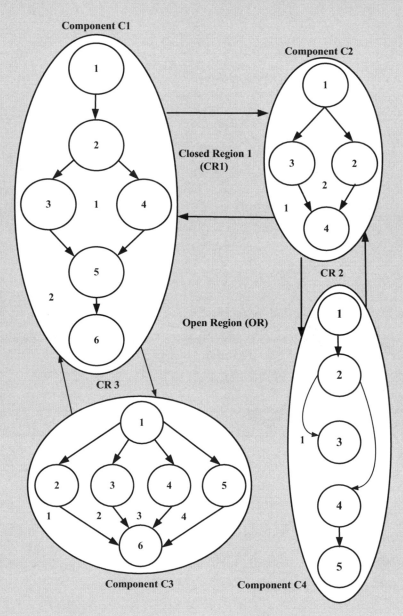

FIGURE 5.6
Interaction flow graph of four components with 3 CRs.

Method 1:
From Figure 5.6, it is noted that,
$|E| = 28$, $|V| = 21$, and $|P| = 4$, therefore,

$$V(G) = |E| - |V| + 2 + |P|$$

$$= 28 - 21 + 2 + 4 = 13$$

Method 2:
From Figure 5.6, it is noted that,

$n = 4$, $m = 3$, $(IC)_1 = 2$, $(IC)_2 = 2$, $(IC)_3 = 2$, $(IC)_4 = 1$, $CR = 3$, $OR = 1$, therefore,

$$V(G) = \sum_{i=1}^{n} (IC)_i + \sum_{j=1}^{m} (CR)_j + OR$$

$$V(G) = (2 + 2 + 4 + 1) + 3 + 1 = 13$$

SCENARIO 5: INTERACTION SCENARIO AMONG FOUR COMPONENTS AND 7 CRs

Figure 5.7 shows an interaction scenario between four components. The internal structures of components are also shown in the figure.

Method 1:
From Figure 5.7, it is noted that,
$|E| = 32$, $|V| = 21$, and $|P| = 4$, therefore,

$$V(G) = |E| - |V| + 2 + |P|$$

$$= 32 - 21 + 2 + 4 = 17$$

Method 2:
From Figure 5.7, it is noted that,

$n = 4$, $m = 7$, $(IC)_1 = 2$, $(IC)_2 = 2$, $(IC)_3 = 4$, $(IC)_4 = 1$, $CR = 7$, $OR = 1$, therefore,

$$V(G) = \sum_{i=1}^{n} (IC)_i + \sum_{j=1}^{m} (CR)_j + OR$$

$$V(G) = (2 + 2 + 4 + 1) + 7 + 1 = 17$$

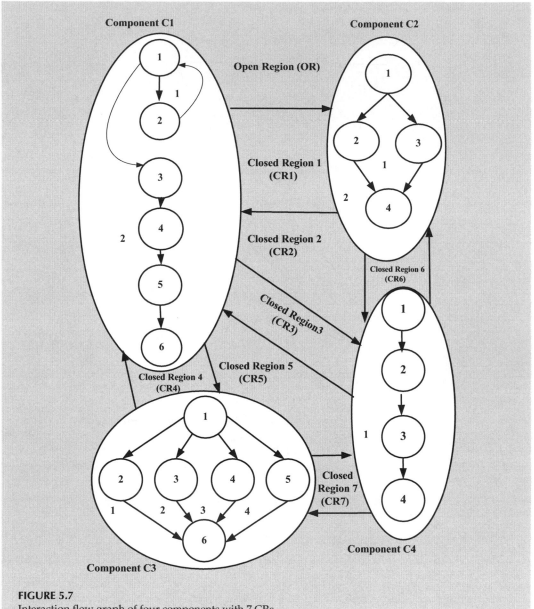

FIGURE 5.7
Interaction flow graph of four components with 7 CRs.

SCENARIO 6: INTERACTION SCENARIO AMONG FIVE COMPONENTS AND 10 CRs

Figure 5.8 shows an interaction scenario between four components. The internal structures of components are also shown in the figure.

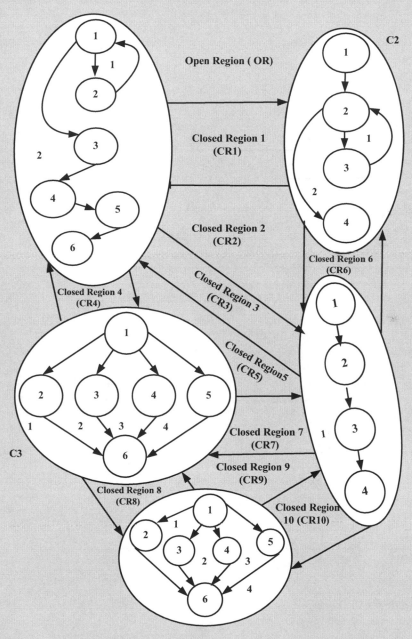

FIGURE 5.8
Interaction flow graph of five components with 10 CRs.

Method 1:
From Figure 5.8, it is noted that,
$|E| = 44$, $|V| = 27$, and $|P| = 5$, therefore,

$$V(G) = |E| - |V| + 2 + |P|$$

$$= 44 - 27 + 2 + 5 = 24$$

Method 2:
From Figure 5.8, it is noted that,

$n = 5$, $m = 10$, $(IC)_1 = 2$, $(IC)_2 = 2$, $(IC)_3 = 4$, $(IC)_4 = 1$, $(IC)_5 = 4$, $CR = 10$, $OR = 1$,

therefore,

$$V(G) = \sum_{i=1}^{n} (IC)_i + \sum_{j=1}^{m} (CR)_j + OR$$

$$V(G) = (2 + 2 + 4 + 1 + 4) + 10 + 1 = 24$$

Summary

In the context of software engineering, complexity is probably the most crucial factor of software design. The emphasis for researchers is on devising methods and techniques that help to reduce the overall complexity of the software. Complexity is expressed as the estimation of efficient use of resources and the level of difficulty to handle the software. Component-based software applications are composed of independently deployable components. Assembling tthese components has a common intesion to contribute their functionalities to the system.

This chapter focuses on three major complexity computation metrics: in-out interaction complexity, integration complexity and cyclomatic complexity for component-based software. Complexity graphs have been used to define these metrics.

The in-out interaction is defined on the basis of edges that are coming towards to the component as well as edges leaving the component. Six different but relative metrics for interactions among components are defined:

1. Total interactions of a component
2. Total interactions of component-based software
3. Interaction ratio of a component
4. Interaction ratio of component-based software
5. Average interactions among components
6. Interaction percentage of component

Case studies explain the defined metrics.

Finally, the concept of cyclomatic complexity for component-based software is discussed, using control-flow graph notation. A number of case studies having a varied number of components are used to elaborate cyclomatic complexity. The given cyclomatic complexity is also compared with McCabe's cyclomatic complexity, which is defined in the context of programs and small-scale software.

References

Bauer, F. et al. 1968. *Software Engineering: A Report on a Conference Sponsored by NATO Science Committee.* Scientific Affairs Division, NATO, Brussels.

Boehm, B. W. 1981. *Software Engineering Economics.* Prentice Hall, Englewood Cliffs, NJ.

Gaedke, M. and J. Rehse. 2000. "Supporting Compositional Reuse in Component-Based Web Engineering." In *Proceedings of ACM symposium on Applied computing (SAC '00).* ACM Press, New York, 927–933.

IEEE. 1987. *Software Engineering Standards.* IEEE Press, New York.

IEEE Standard 610.12-1990. 1990. *Glossary of Software Engineering Terminology.* IEEE, New York, ISBN: 1–55937–079–3 .

Krueger, C. W. 1992. "Software Reuse." *ACM Computing Surveys (CSUR),* 24(2): 131–183.

McCabe, T. 1976. "A Complexity Measure." *IEEE Transactions on Software Engineering,* 2(8): 308–320.

Pressman S. R. 2005. *Software Engineering A Practitioner's Approach,* 6th edn. TMH International Edition, New York.

Schach, S. 1990. "Software Engineering." Vanderbilt University, Aksen Association.

Shepperd, M. 1988. "A Critique of Cyclomatic Complexity as Software Metric." *Software Engineering Journal,* 3(2): 30–36.

Tiwari, U. and S. Kumar. 2014. "Cyclomatic Complexity Metric for Component Based Software." *ACM SIGSOFT Software Engineering Notes,* 39(1): 1–6.

6

Component-Based Software Testing Techniques and Test-Case Generation Methods

6.1 Introduction

> Testing is the process of executing a program with the intent of finding errors
>
> **(Myers 1979)**

Testing is one of the core activities of the software development process. In component-based development the approach emphasizes the "use of pre-built and pre-tested components." Here the focus of developers is on black-box testing (functional testing) as well as white-box testing (structural testing). Black-box testing emphasizes the behavioral attributes of the components when they interact with each other. White-box testing techniques are used to address the testing of the structural design and internal code of the software.

In this chapter, functional testing and structural testing strategies, and test-case generation techniques for CBSE are discussed. When components are integrated, they produce explicit effects called integration effects. Integration-effect methodology is a black-box technique as it covers the input and output domains only. In this chapter black-box and white-box test-case generation methods are discussed. These methods are compared only with the boundary-value analysis method since other black-box techniques require specific input conditions and the number of test cases depends on those conditions. A white-box testing technique named cyclomatic complexity is described.

6.2 Testing Techniques

For testing purposes, software constructs are divided into two broad categories: *input/output constructs* and *process/logic/code constructs*. Input/output constructs refer to inputs provided to the software and outputs produced by it. Normally input/output is presented in the form of data, information and specific values. Process/logic/code constructs refer to the actual processing of software, including coding and internal structure. Testing techniques are classified into two major classes according to the construct type:

1. Black-box testing
2. White-box testing

6.2.1 Black-Box Testing

Black-box testing techniques only consider software inputs and outputs. These techniques apply to software whose code is not available or accessible. Input and output domains are divided into various classes or partitions and tested separately. Black-box emphasizes the external behavioral attributes of the components when they interact with each other. An overview of black-box testing is shown in Figure 6.1. Tester control is on input and output domains only. The internal structure and logic of the program and software are not accessible. The literature describes various categories of black-box testing techniques: boundary-value analysis, equivalence-class partitioning, decision table-based and cause-effect graphing (see Chapter 3).

Input to a program/software is given in the form of data, information or specific values as per the requirements. After processing, the achieved outputs are collected and tested. Since the logic is hidden from the testers, they only control the inputs/outputs of the software.

6.2.2 White-Box Testing

White-box testing techniques consider the internal logic of the software and apply to its structural code. In this technique the program statements are checked and errors are fixed. White-box testing techniques are used to address the testing of the structural design and internal code of the software. An overview of white-box testing methods is shown in Figure 6.2. White-box testing not only tests the structural behavior of software but also considers the control flows, basis paths, data structures (within sub-component and outside the sub-components), independent paths, logical mistakes, coding errors, incoming and outgoing interfaces, and semantic errors. Here input provided to the software is assumed to be true and the functionalities provided by the software are checked (see Chapter 3).

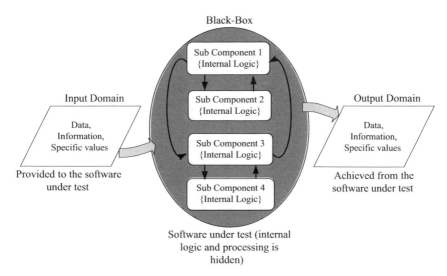

FIGURE 6.1
Black-box component testing.

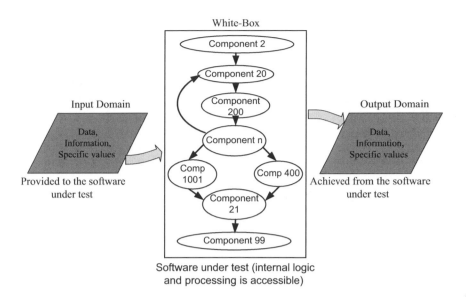

FIGURE 6.2
White-box component testing.

6.3 Testing Levels for Component-Based Software

The fundamental concept behind component-based software is to develop new software constructs by reusing available application constructs rather than developing them from the first to the last line. Component-based software promises savings in overall development cost by reusing components, and in terms of development effort and testing.

The basic difficulty with component-based software testing is that, at the time when pre-built, pre-tested and error-free components are integrated, they may fail to provide the intended and expected function. Off-the-shelf components may not satisfy the requirements of the application and may be incompatible with existing components, having been developed for other purposes, and they may have varied usage contexts. Available testing methodologies may facilitate validation of the interconnectivity of connected components, but they do not provide clear and defined levels of testing in the context of components and levels of component-based software. Software developers may therefore be facing testing problems in CBS development, especially in the distributed environment.

Although conventional development testing techniques may help to promise the quality of individual components, they cannot guarantee the quality assurance of component-based applications.

It is clear that testing methods for component-based systems are different from those used in traditionally developed systems, both conceptually and in terms of their applicability. Component-based software testing includes each and every action concerned with the testing of specific components and also the test development activities of these components. It also includes testing activities during engineering of applications in early stages of development (Gross 2009).

Typically, components are developed and used in various environments and contexts. The component owner/provider, who can assume only limited use scenarios for the component, cannot test it in every possible context (Weyuker 1998, 2001). Normally components are provided as black-box, so their internal code is generally not accessible (Weyuker 1998, Harrold et al. 1999), reducing control and testability of components.

Testing levels of component-based software must integrate software testing methods into a predefined and planned sequence of actions. Testing methodologies must incorporate planning of testing, test-case design, implementation of testing, testing results and the assessment of collected data. A component is typically made up of different independent sub-components from various contexts. Therefore, a testing technique that addresses these components not only individually but as a whole is required. Conventional testing methodologies are applicable to small-scale programs and software, but they appear inefficient when applied in the context of component-based software.

6.3.1 Component Testing Strategy at Individual Component Level

Component-based software integrates components using well-specified and properly implemented interfaces. These compatible interfaces provide the platform for the assembling of individual components. The testing of interfaces ensures that communicated information and data are properly conveyed to/from the components. The internal data residing in these components are inspected for correctness. Boundary logics and boundary values are checked to guarantee that the component works at its extreme values. Every independent path that represents the independent logics of the component should be tested to ensure the reliability and correctness of the component. During the testing of individual components, syntactical faults, semantic errors, logical errors and coding mistakes are checked before any other test is initiated. In addition, independent paths, interfaces related to in-and-out interactions, local data structures (within and outside sub-components), boundary conditions and error handling paths are tested during component testing. All these tests must be backed up and recorded by test cases.

6.3.2 Test Specification

Every test has its attributes and characteristics. A test specification defines and specifies the attributes of a test. It describes each individual test, the cardinality and the cardinality ratio of test items belonging to each domain, the test elements and their formats, the testing process for each independent test item, test-case recording methods, test-case semantics, testing dates, and the minimum and maximum time taken to conduct the tests. In general, a test specification provides the complete structure of the testing elements and its corresponding attributes.

6.3.3 Test Plan

The software under consideration has its own requirements and testers try to verify the software in accordance with those requirements. A test plan is like a manuscript directive that addresses the verification of the particular software's requirement and the design specifications. Testers, developers and sometimes end users contribute inputs to the test plan, which consists of:

- Testing process
- Basic inputs and final outcomes
- Software elements that should be tested
- Required quality levels
- People involved in the testing process, i.e., tester, developer and end user

6.3.4 Test Cases

In addition to other elements, the component-based test case is a complete document that contains three basic constructs:

- Actual inputs to the system
- Actual outputs achieved from the system
- Testing logic through which input is transformed into output

A test case executes the test plans, records the actual outcomes and compares them with the expected ones. Test cases are the documents that are used at various levels of testing.

6.3.5 Test Documentation

During component-level testing the major focus is on input, output, and internal as well as related and supporting logics of the component (Figure 6.3). Test documentation is a testing repository not only for the individual component but for the complete testing data of component-based software during integration and system-level testing.

- **Syntactical faults:** In component-based software, components interact through parameters. These parameters are passed through the component's language. The communicating syntax used between different components must be unambiguous and clearly defined. In this communication, interfaces are used to make compatibility between different components that have diverse syntactical problems.
- **Semantic errors:** A semantic error means ambiguity or unclear specification interpreted by the system or the user, which may cause undetermined behavior, incorrect output or no output.
- **Logical errors:** A logical error denotes faulty implementation of algorithms, process logics and similar mistakes that cause incorrect operation. Logical errors are normally not reflected in the form of errors, but they produce wrong solutions.
- **Coding mistakes:** May be a syntax error, semantic error, logical error or a combination of these.
- **Identifying and testing independent paths:** An *independent path* is a component's execution path from source to termination. It includes any new statement at component level or a new component at system level. Generally, the execution of a component or component-based software follows a predefined path. Each independent path designed and defined in the component's structure must be tested to identify and test the execution path.
- **In-out interfaces:** In component-based software, interfaces are used by components, and these interfaces provide data and information to each component. So, in this testing input and output data related to each component are verified.
- **Local data structures (within and outside sub-components):** In in-out interface testing, inter-component data are tested to verify the behavior of components, whereas in local data structure testing, the focus is on testing intra-component data.
- **Boundary conditions:** Logical conditions and the components normally fail at their boundaries. Testers therefore try to check these boundary conditions.
- **Error handling paths:** Components execute according to a pre-specified path, which must be checked for both normal and abnormal execution paths.

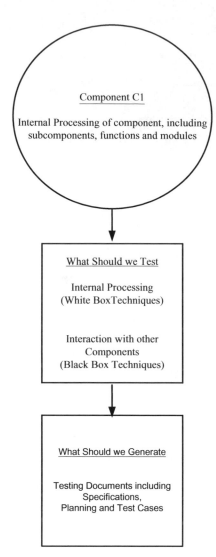

FIGURE 6.3
Testing of individual components. (Adapted from Umesh and Kumar 2017.)

6.4 Test-Case Generation Technique for Black-Box Components

This section discusses a test-case generation technique for black-box components in a component-based software environment. The method is called an integration-effect graph, the principle being that if two or more components are integrated, they will produce an effect. In this technique, an integration-effect matrix is constructed which counts the total number of test cases for complex and large component-based software. This is a black-box technique as it is only concerned with the external behavior of components. The internal structure of components is not taken into account during the test-case count.

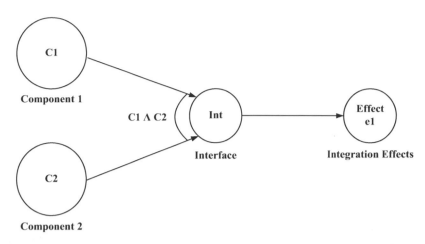

FIGURE 6.4
Integration-effect graph of two components.

6.4.1 Component Integration-Effect Graph

This technique uses the architecture design of the developed software to draw an integration-effect graph. Components are assembled to communicate and share data, control or information (Basili 2001, Chen 2011). Component assembly is according to the design of the software (Gill and Balkishan 2008), as is their effect. This effect graph uses nodes or vertices to represent components, effects and interfaces that are used to integrate the components. Integration among components is represented as edges. The integration-effect graph in Figure 6.4 shows the integration effect of two components, C1 and C2. The integration operation is denoted C1 ∧ C2.

6.4.2 Test-Case Generation Using Integration-Effect Graph

The integration-effect graph shows the structure of integrations and the integration-effect matrix shows their effect. The matrix is made up of columns and rows representing components and integration effects respectively, as shown in Table 6.1.

The integration-effect matrix consists of two categories of values:

1. *Value of integration among components, and*
2. *Value of integration effect, denoted as int-eff, which is produced because of effect of integration, if any.*

The integration-effect values are evaluated using the following formula, as:

Components Effect ∧ Integration effect produced by integrating components

TABLE 6.1

Probable Values of Integration-Effect Matrix of Two Components

Components	C1	Integration Effect	C2	Integration Effect
C1	1	Effect of (C1): 0/1	0: if not connected 1: if connected	Integration effect of (C1∧C2): 0/1
C2	0: if not connected 1: if connected	Integration effect of (C2∧C1): 0/1	1	Effect of (C2): 0/1

where these integration effects are Boolean values, either 0 or 1.

This formula is used to construct the integration-effect matrix, which counts test cases. Producing test cases is a two-step activity:

Step 1. Record the effect of integration of components. If the effect is "1," it will contribute to the count of test cases.

Step 2. Count the "1s" recorded in Step1.

The following case studies with different scenarios illustrate the method described above.

SCENARIO 1: INTEGRATION EFFECT BETWEEN TWO COMPONENTS

Figure 6.1 shows a scenario with integration between two components. Integration between components C1 and C2 takes place with the help of an interface which is compatible with both the components. The integration effect is shown as an "Effects" node. When the integration effects are computed, both the effects of contributing components and those of integrations are taken into account. The effect of integration can be computed as

$$\text{Integration of } (C1 \wedge C2) = \text{Effect of } (C1) \wedge \text{ Effect of } (C2) \wedge$$
$$\text{Integration effect of } (C1 \wedge C2)$$

where

Integration effect of (C1 $\wedge$ C2) denotes the integration of components,
Effect of (C1) denotes the effects in the software produced by C1,
Effect of (C2) denotes the effects in the software produced by C2,
Integration effect of (C1 $\wedge$ C2) denotes the effects produced by interaction of components C1 and C2, and

$\wedge$ shows the "AND" operation, Effect of (Ci) is denoted "1" if Ci contains no faults, but is denoted "0" if Effect of (Ci) contains any error or fault. Note that the integration value of (C1 $\wedge$ C2) in the integration-effect matrix will be 1 if and only if Effect of (C1) = 1, Effect of (C2) = 1, and Integration effect of (C1 $\wedge$ C2) = 1. If any of the given effects are 0, the integration value of (C1 $\wedge$ C2) will be 0. That is,

- Integration of (C1 $\wedge$ C2) = 1 only if Effect of (C1) = 1, Effect of (C2) = 1, and Integration effect of (C1 $\wedge$ C2) = 1.
- Integration of (C1 $\wedge$ C2) = 0 if Effect of (C1) = 0, Effect of (C2) = 0, and Integration effect of (C1 $\wedge$ C2) = 0.
- Integration of (C1 $\wedge$ C2) = 0 if Effect of (C1) = 0, Effect of (C2) = 0, and Integration effect of (C1 $\wedge$ C2) = 1.
- Integration of (C1 $\wedge$ C2) = 0 only if Effect of (C1) = 0, Effect of (C2) = 1, and Integration effect of (C1 $\wedge$ C2) = 0.
- Integration of (C1 $\wedge$ C2) = 0 only if Effect of (C1) = 1, Effect of (C2) = 0, and Integration effect of (C1 $\wedge$ C2) = 0.
- Integration of (C1 $\wedge$ C2) = 0 only if Effect of (C1) = 1, Effect of (C2) = 1, and Integration effect of (C1 $\wedge$ C2) = 0.

- Integration of (C1 ∧ C2) = 0 only if Effect of (C1) = 1, Effect of (C2) = 0, and Integration effect of (C1 ∧ C2) = 1.
- Integration of (C1 ∧ C2) = 0 only if Effect of (C1) = 0, Effect of (C2) = 1, and Integration effect of (C1 ∧ C2) = 1.

Integration-Effect Matrix of Two Components

The next stage is to construct the integration-effect matrix, consisting of columns and rows. The columns represent components and the rows represent effects. Components that are connected directly contain "1" in the matrix, that is, a direct edge between two components in the integration-effect graph is represented as 1 in the integration-effect matrix. It is represented as 0 if there is no edge between two components. For example, in Figure 6.3, there is a direct edge between components C1 and C2, which is represented as 1 in the integration-effect matrix in Table 6.1. Every component is assumed to be directly connected by itself, therefore it is represented as 1 in the matrix. Component C1 is cohesive with itself; hence C1 to C1 is denoted 1 in the matrix.

The probable values that can occur in the integration-effect matrix are defined in Table 6.1.

Integration-Effect Matrix for Fault-Free Components

When there is no fault in the components, the integration effect will also be fault free. In the integration-effect matrix this is denoted 1. Table 6.2 shows the integration-effect matrix when there is no error or fault in the components and their integration.

In the first row, that is, in row C1:

Effect of (C1) = 1, meaning component C1 is fault free,

Effect of (C2) = 1, meaning component C2 is fault free,

Integration effect of (C1 ∧ C2) = 1, meaning interaction between C1 and C2 is fault free.

Similarly, in the second row, that is in row C2:

Effect of (C1) = 1, meaning component C1 is fault free,

Effect of (C2) = 1, meaning component C2 is fault free,

Integration effect of (C1 ∧ C2) = 1, meaning interaction between C1 and C2 is fault free.

TABLE 6.2

Integration Effect Matrix of Two Components Without Faults

Components	C1	Integration-Effect	C2	Integration-Effect
C1	1	1	1	1
C2	1	1	1	1

Integration-Effect Matrix with Faulty Components

When components contain any fault, their integration will also contain errors. Table 6.3 represents the integration-effect matrix when a component or integration between components contains a fault. At the place where components contain any kind of error or fault, 0 is positioned. For example, the integration effect of component C1 and C2 is faulty, hence the value of Integration effect of (C1 ∧ C2) = 0.

In the first row, that is in row C1:

Effect of (C1) = 1, meaning component C1 is fault free,

Effect of (C2) = 1, meaning component C2 is fault free,

Integration effect of (C1 ∧ C2) = 1, meaning interaction between C1 and C2 is fault free.

In the second row, that is in row C2:

Effect of (C1) = 1, meaning component C1 is fault free,

Effect of (C2) = 1, meaning component C2 is fault free,

Integration effect of (C1 ∧ C2) = 0, meaning interaction between C1 and C2 is faulty.

Computation of Test Cases from Integration-Effect Matrix

To compute the test cases from the integration-effect matrix, we have to calculate the number of test cases where an individual component is involved. It is calculated as follows:

Test cases for component C1, that is the case when C1 is involved

$$= (\text{Count of 1s below "Integration Effect" column in the first row}) - 1 = 2 - 1 = 1$$

TABLE 6.3

Integration Effect Matrix of Two Components with Faults

Components	C1	Integration-Effect	C2	Integration-Effect
C1	1	1	1	1
C2	1	0	1	1

TABLE 6.4

Integration Effect Matrix of Two Components

Components	C1	Integration-Effect	C2	Integration-Effect
C1	1	1	1	1
C2	1	1	1	1

Test cases for component C2, that is the case when C2 is involved

$$= \left(\text{Count of 1s below "Integration Effect" column in the second row}\right) - 1 = 2 - 1 = 1$$

Overall test cases produced using integration-effect matrix:

= Count of test cases for first component C1 + Count of test cases for first component C2

$$= 1 + 1 = 2.$$

When we apply the boundary-value analysis (BVA) technique to this scenario we get:
If there are n number of components in the software, the least amount of test cases produced by BVA are 4n + 1.
In the given scenario we have n = 2,
Then, count of test cases = 4 * n + 1

$$= 4 * 2 + 1 = 9$$

SCENARIO 2:   INTEGRATION EFFECT BETWEEN THREE COMPONENTS

In this scenario we have three components interacting with each other as shown in Figure 6.5. This figure depicts following connections:

C1 is making direct connection with components C2 and C3.
C3 is directly linked with C1 only.
C2 is directly connected with C1 only.

There is no direct link between C2 and C3.

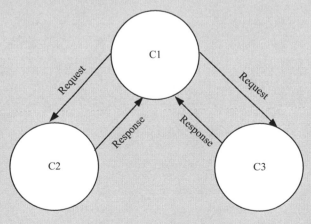

FIGURE 6.5
Interactions among three components.

Integration-Effect Graph for Three Components

The integration-effect graph for the integration structure shown in Figure 6.5 is given in Figure 6.6. From Figure 6.5, we can see that at any point in time, only two components are attached via the interface. The integration effects are recorded as "e1" and "e2" respectively.

The integration effect can be computed as:

$$\text{Integration of} \left(C1 \wedge C2 \wedge C3 \right)$$

$$= \text{Effect of} \left(C1 \right) \wedge \text{ Effect of} \left(C2 \right) \wedge \text{Effect of} \left(C3 \right) \wedge \text{Integration effect of} \left(C1 \wedge C2 \right) \wedge$$
$$\text{Integration effect of} \left(C2 \wedge C3 \right) \wedge \text{Integration effect of} \left(C1 \wedge C3 \right)$$

where

Integration of (C1 ∧ C2 ∧ C3) denotes the integration of 3 given components,
Effect of (C1) denotes the effects in the software produced by C1; it is either 0 or 1
Effect of (C2) denotes the effects in the software produced by C2; it is either 0 or 1
Effect of (C3) denotes the effects in the software produced by C3; it is either 0 or 1

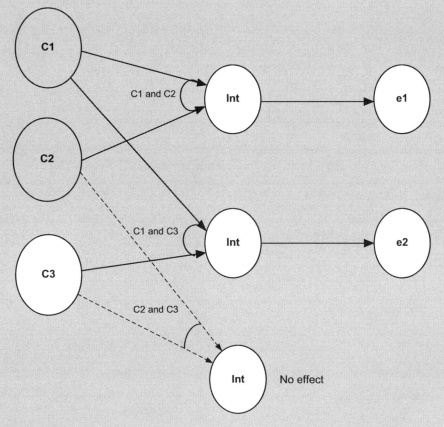

FIGURE 6.6
Integration-effect graph of three components.

Integration effect of (C1 ∧ C2) denotes the effect produced by interaction of components C1 and C2; it is either 0 or 1.

Integration effect of (C2 ∧ C3) denotes the effect produced by interaction of components C2 and C3; it is either 0 or 1.

Integration effect of (C1 ∧ C3) denotes the effect produced by interaction of components C1 and C3; it is either 0 or 1.

∧ shows the "AND" operation.

Effect of (C1, C2, and C3) is denoted "1" if components contain no faults, but "0" if Effect of (C1, C2, and C3) contains any error or fault. The integration value of (C1 ∧ C2 ∧ C3) in the integration-effect matrix will be 1 if and only if Effect of (C1) = 1, Effect of (C2) = 1, Effect of (C3) = 1, and Integration effect of (C1 ∧ C2 ∧ C3) = 1. If any of the given effects is 0, the integration value of (C1 ∧ C2 ∧ C3) will be 0.

Integration-Effect Matrix for Three Components

The probable integration-effect matrix for 3 components is shown in Table 6.5.
Table 6.6 presents the integration-effect matrix for the above scenario (Figure 6.5).

Computation of Test Cases from Integration-Effect Matrix

To compute the test cases from the integration-effect matrix, we have to calculate the number of test cases where an individual component is involved. It is calculated as:

Test cases for component C1, that is the case when C1 is involved

$$= \left(\text{Count of 1s below "Integration Effect" column in the first row} \right) - 1 = 3 - 1 = 2$$

Test cases for component C2, that is the case when C2 is involved

$$= \left(\text{Count of 1s below "Integration Effect" column in the second row} \right) - 1 = 2 - 1 = 1$$

TABLE 6.5

Probable Values of Integration-Effect Matrix of Three Components

Components	C1	Integration Effect	C2	Integration Effect	C3	Integration Effect
C1	1	Effect of (C1): 0/1	0 or 1	Integration effect of (C1 ∧ C2): 0/1	0 or 1	Integration effect of (C1∧ C3): 0/1
C2	0 or 1	Integration effect of (C2∧C1): 0/1	1	Effect of (C2): 0/1	0 or 1	Integration effect of (C2 ∧ C3): 0/1
C3	0 or 1	Integration effect of (C3 ∧ C1): 0/1	0 or 1	Integration effect of (C3 ∧ C2): 0/1	1	Effect of (C3): 0/1

TABLE 6.6

Actual Values of Integration-Effect Matrix of Three Components

Components	C1	Integration Effect	C2	Integration Effect	C3	Integration Effect
C1	1	1	1	1	1	1
C2	1	1	1	1	0	0
C3	1	1	0	0	1	1

Test cases for component C3, that is the case when C3 is involved

$$= \left(\text{Count of 1s below "Integration Effect" column in the third row} \right) - 1 = 2 - 1 = 1$$

Overall test cases produced using integration-effect matrix:

= Count of test cases for first component C1 + Count of test cases for first component C2 + Count of test cases for first component C3 = 2 + 1 + 1 = 4.

When we apply boundary-value analysis techniques to this scenario we get:
If there are n number of components in the software, the least amount of test cases produced by BVA are 4n + 1.
In the given scenario we have n = 3,
Then, count of test cases = 4 * 3 + 1

$$= 4 * 3 + 1 = 13$$

SCENARIO 3:   INTEGRATION EFFECT BETWEEN FOUR COMPONENTS

In this scenario we have four components interacting with each other (Figure 6.7). The following connections occur:

C1 is making direct connection with components C2 and C3.
C2 is directly linked with C1 and C4.
C3 is directly connected with C1 and C4.
C4 is attached to C2 and C3.

Integration-Effect Graph for Three Components

This is a four-component integration scenario. The integration-effect graph according to the integration structure shown in Figure 6.7 is given in Figure 6.8.

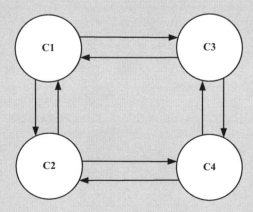

FIGURE 6.7
Interactions among four components.

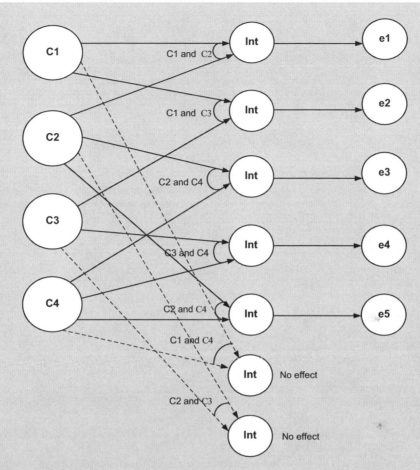

FIGURE 6.8
Integration-effect graph of four components.

The components interact and produce effects, recorded as "e1," "e2," "e3," "e4" and "e5" respectively.

The effect of integration can be computed as:

Integration of $(C1 \wedge C2 \wedge C3 \wedge C4)$

$= $ Effect of $(C1) \wedge$ Effect of $(C2) \wedge$ Effect of $(C3) \wedge$ Effect of $(C4) \wedge$
    Integration effect of $(C1 \wedge C2) \wedge$ Integration effect of $(C1 \wedge C3)$

$\wedge$ Integration effect of $(C1 \wedge C4) \wedge$ Integration effect of $(C2 \wedge C1)$

$\wedge$ Integration effect of $(C2 \wedge C3) \wedge$ Integration effect of $(C2 \wedge C4)$

$\wedge$ Integration effect of $(C3 \wedge C1) \wedge$ Integration effect of $(C3 \wedge C2)$

$\wedge$ Integration effect of $(C3 \wedge C4) \wedge$ Integration effect of $(C4 \wedge C1)$

$\wedge$ Integration effect of $(C4 \wedge C2) \wedge$ Integration effect of $(C4 \wedge C3)$

where

Integration of (C1 $\wedge$ C2 $\wedge$ C3 $\wedge$ C4) denotes the integration of four given components.

Effect of (C1) denotes the effects in the software produced by C1; it is either 0 or 1.

Effect of (C2) denotes the effects in the software produced by C2; it is either 0 or 1.

Effect of (C3) denotes the effects in the software produced by C3; it is either 0 or 1.

Effect of (C4) denotes the effects in the software produced by C4; it is either 0 or 1.

Integration effect of (C1 $\wedge$ C2) denotes the effects produced by interaction of components C1 and C2; it is either 0 or 1.

Integration effect of (C1 $\wedge$ C3) denotes the effects produced by interaction of components C1 and C3; it is either 0 or 1.

Integration effect of (C1 $\wedge$ C4) denotes the effects produced by interaction of components C1 and C4; it is either 0 or 1.

Integration effect of (C2 $\wedge$ C1) denotes the effects produced by interaction of components C2 and C1; it is either 0 or 1.

Integration effect of (C2 $\wedge$ C3) denotes the effects produced by interaction of components C2 and C3; it is either 0 or 1.

Integration effect of (C2 $\wedge$ C4) denotes the effects produced by interaction of components C2 and C4; it is either 0 or 1.

Integration effect of (C3 $\wedge$ C1) denotes the effects produced by interaction of components C3 and C1; it is either 0 or 1.

Integration effect of (C3 $\wedge$ C2) denotes the effects produced by interaction of components C3 and C2; it is either 0 or 1.

Integration effect of (C3 $\wedge$ C4) denotes the effects produced by interaction of components C3 and C4; it is either 0 or 1.

Integration effect of (C4 $\wedge$ C1) denotes the effects produced by interaction of components C4 and C1; it is either 0 or 1.

Integration effect of (C4 $\wedge$ C2) denotes the effects produced by interaction of components C4 and C2; it is either 0 or 1.

Integration effect of (C4 $\wedge$ C3) denotes the effects produced by interaction of components C4 and C3; it is either 0 or 1.

$\wedge$ shows the "AND" operation.

Effect of (C1, C2, C3 and C4) is denoted "1" if components contain no faults, but "0" if Effect of (C1, C2, C3 and C4) contains any error or fault. The integration value of (C1 $\wedge$ C2 $\wedge$ C3 $\wedge$ C4) in the integration-effect matrix will be 1 if and only if Effect of (C1) = 1, Effect of (C2) = 1, Effect of (C3) = 1, Effect of (C4) = 1, and Integration effect of (C1 $\wedge$ C2 $\wedge$ C3 $\wedge$ C4) = 1. If any of the given effects is 0, the integration value of (C1 $\wedge$ C2 $\wedge$ C3 $\wedge$ C4) will be 0.

Integration-Effect Matrix of Four Components

A probable integration-effect matrix for four components is shown in Table 6.7.

Table 6.8 presents the integration-effect matrix for the above scenario (Figure 6.8).

TABLE 6.7

Probable Values of Integration-Effect Matrix of Four Components

Components	C1	Integration Effect	C2	Integration Effect	C3	Integration Effect	C4	Integration Effect
C1	1	Effect of (C1): 0/1	1	Integration effect of (C1 ∧ C2): 0/1	0 or 1	Integration effect of (C1 ∧ C3): 0/1	0 or 1	Integration effect of (C1 ∧ C4): 0/1
C2	0 or 1	Integration effect of (C2 ∧ C1): 0/1	1	Effect of (C2): 0/1	0 or 1	Integration effect of (C2 ∧ C3): 0/1	0 or 1	Integration effect of (C2 ∧ C4): 0/1
C3	0 or 1	Integration effect of (C3 ∧ C1): 0/1	0 or 1	Integration effect of (C3 ∧ C2): 0/1	1	Effect of (C3): 0/1	0 or 1	Integration effect of (C3 ∧ C4): 0/1
C4	0 or 1	Integration effect of (C4 ∧ C1): 0/1	0 or 1	Integration effect of (C4 ∧ C2): 0/1	0 or 1	Integration effect of (C4 ∧ C3): 0/1	1	Effect of (C4): 0/1

TABLE 6.8

Actual Values of Integration-Effect Matrix of four Components

Components	C1	Effect	C2	Effect	C3	Effect	C4	Effect
C1	1	1	1	1	1	1	1	1
C2	1	1	1	1	0	0	1	1
C3	1	1	0	0	1	1	1	1
C4	1	1	1	1	1	1	1	1

Computing Test Cases from Integration-Effect Matrix

To compute the test cases from the integration-effect matrix, we have to calculate the number of test cases where individual components are involved. It is calculated as:

Test cases for component C1, that is the case when C1 is involved

$$= \left(\text{Count of 1s below "Integration Effect" column in the first row} \right) - 1 = 4 - 1 = 3$$

Test cases for component C2, that is the case when C2 is involved

$$= \left(\text{Count of 1s below "Integration Effect" column in the second row} \right) - 1 = 3 - 1 = 2$$

Test cases for component C3, that is the case when C3 is involved

$$= \left(\text{Count of 1s below "Integration Effect" column in the third row} \right) - 1 = 3 - 1 = 2$$

Test cases for component C4, that is the case when C4 is involved

$$= \left(\text{Count of 1s below "Integration Effect" column in the fourth row} \right) - 1 = 4 - 1 = 3$$

Overall test cases produced using integration-effect matrix:

= Count of test cases for first component C1 + Count of test cases for first
    component C2 + Count of test cases for first component C3
    + Count of test cases for first component C3.
= 3 + 2 + 2 + 3 = 10.

When we apply the boundary-value analysis technique to this scenario we get:
If there are n number of components in the software, the least amount of test cases
produced by BVA are 4n + 1.
In the given scenario we have n = 4,
Then, count of test cases = 4 * 4 + 1

    = 4 * 4 + 1 = 17

SCENARIO 4: INTEGRATION EFFECT BETWEEN FIVE COMPONENTS

In this scenario we have five components interacting with each other (Figure 6.9).
The following connections occur:

C1 is making a direct connection with components C2, C3 and C4.
C2 is directly linked with C1, C3 and C4.
C3 is directly connected with C1, C2, C4 and C5.
C4 is attached to C1, C2, C3 and C5.
C5 is attached to C3 and C4.

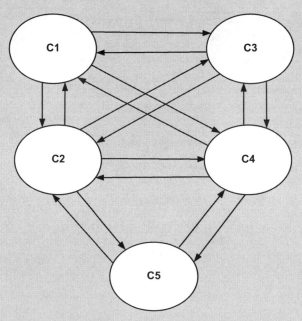

FIGURE 6.9
Interactions among five components.

Integration-Effect Graph for Three Components

This is a five-component integration scenario. The integration-effect graph according to the integration structure shown in Figure 6.9 is given in Figure 6.10.

Components interact and produce effects. The effects of integrations are recorded as "e1," "e2," "e3," "e4," "e5," "e6," "e7" and "e8," respectively.

The effect of integration can be computed as:

Integration of $(C1 \wedge C2 \wedge C3 \wedge C4 \wedge C5)$

$= \text{Effect of} (C1) \wedge \text{Effect of} (C2) \wedge \text{Effect of} (C3) \wedge \text{Effect of} (C4) \wedge \text{Effect of} (C5) \wedge$
Integration effect of $(C1 \wedge C2) \wedge$ Integration effect of $(C1 \wedge C3) \wedge$ Integration effect of $(C1 \wedge C4) \wedge$ Integation effect of $(C1 \wedge C5) \wedge$ Integration effect of $(C2 \wedge C1) \wedge$ Integration effect of $(C2 \wedge C3) \wedge$ Integration effect of $(C2 \wedge C4) \wedge$ Integration

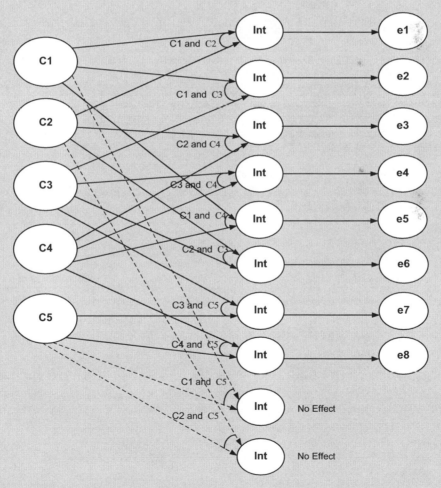

FIGURE 6.10
Integration-effect graph of five components.

effect of (C2 $\wedge$ C5) Integration effect of (C3 $\wedge$ C1) Integration effect of (C3 $\wedge$ C2)
$\wedge$ Integration effect of (C3 $\wedge$ C4) $\wedge$ Integation effect of (C3 $\wedge$ C5) $\wedge$ Integration
effect of (C4 $\wedge$ C1) $\wedge$ Integration efect of (C4 $\wedge$ C2) $\wedge$ Integration effect of
(C4 $\wedge$ C3) $\wedge$ Integation Effect of (C4 $\wedge$ C5) $\wedge$ Integration effect of $\left(\text{C5} \wedge \text{C1} \right)$
$\wedge$ Integration effect of $\left(\text{C5} \wedge \text{C2} \right)$ $\wedge$ Integration effect of $\left(\text{C5} \wedge \text{C3} \right)$ $\wedge$ Integation
efect of $\left(\text{C5} \wedge \text{C4} \right)$

where

Integration of (C1 $\wedge$ C2 $\wedge$ C3 $\wedge$ C4 $\wedge$ C5) denotes the integration of five given
components,

Effect of (C1) denotes the effects in the software produced by C1; it is either
0 or 1

Effect of (C2) denotes the effects in the software produced by C2; it is either
0 or 1

Effect of (C3) denotes the software effects produced by C3; it is either 0 or 1

Effect of (C4) denotes the effects in the software produced by C4; it is either
0 or 1

Effect of (C4) denotes the effects in the software produced by C5; it is either
0 or 1

Integration effect of (C1 $\wedge$ C2) denotes the effects produced by interaction of
components C1 and C2; it is either 0 or 1.

Integration effect of (C1 $\wedge$ C3) denotes the effects produced by interaction of
components C1 and C3; it is either 0 or 1.

Integration effect of (C1 $\wedge$ C4) denotes the effects produced by interaction of
components C1 and C4; it is either 0 or 1.

Integration effect of (C1 $\wedge$ C5) denotes the effects produced by interaction of
components C1 and C5; it is either 0 or 1.

Integration effect of (C2 $\wedge$ C1) denotes the effects produced by interaction of
components C2 and C1; it is either 0 or 1.

Integration effect of (C2 $\wedge$ C3) denotes the effects produced by interaction of
components C2 and C3; it is either 0 or 1.

Integration effect of (C2 $\wedge$ C4) denotes the effects produced by interaction of
components C2 and C4; it is either 0 or 1.

Integration effect of (C2 $\wedge$ C5) denotes the effects produced by interaction of
components C2 and C5; it is either 0 or 1.

Integration effect of (C3 $\wedge$ C1) denotes the effects produced by interaction of
components C3 and C1; it is either 0 or 1.

Integration effect of (C3 $\wedge$ C2) denotes the effects produced by interaction of
components C3 and C2; it is either 0 or 1.

Integration effect of (C3 $\wedge$ C4) denotes the effects produced by interaction of
components C3 and C4; it is either 0 or 1.

Integration effect of (C3 ∧ C5) denotes the effects produced by interaction of components C3 and C5; it is either 0 or 1.

Integration effect of (C4 ∧ C1) denotes the effects produced by interaction of components C4 and C1; it is either 0 or 1.

Integration effect of (C4 ∧ C2) denotes the effects produced by interaction of components C4 and C2; it is either 0 or 1.

Integration effect of (C4 ∧ C3) denotes the effects produced by interaction of components C4 and C3; it is either 0 or 1.

Integration effect of (C4 ∧ C5) denotes the effects produced by interaction of components C4 and C5; it is either 0 or 1.

Integration effect of (C5 ∧ C1) denotes the effects produced by interaction of components C5 and C1; it is either 0 or 1.

Integration effect of (C5 ∧ C2) denotes the effects produced by interaction of components C5 and C2; it is either 0 or 1.

Integration effect of (C5 ∧ C3) denotes the effects produced by interaction of components C5 and C3; it is either 0 or 1.

Integration effect of (C5 ∧ C4) denotes the effects produced by interaction of components C5 and C4; it is either 0 or 1.

∧ shows the "AND" operation.

Effect of (C1, C2, C3, C4, and C5) is denoted "1" if components contain no faults, but "0" if Effect of (C1, C2, C3, C4 and C5) contains any error or fault. Note that the integration value of (C1 ∧ C2 ∧ C3 ∧ C4 ∧ C5) in the integration-effect matrix will be 1 if and only if Effect of (C1) = 1, Effect of (C2) = 1, Effect of (C3) = 1, Effect of (C4) = 1, Effect of (C5) = 1 and Integration effect of (C1 ∧ C2 ∧ C3 ∧ C4 ∧ C5) = 1. If any of the given effects is 0, the integration value of (C1 ∧ C2 ∧ C3 ∧ C4 ∧ C5) will be 0.

Integration-Effect Matrix for Five Components

A probable integration-effect matrix for four components is shown in Table 6.9.

Table 6.10 presents the integration-effect matrix for the above case (Figure 6.9).

Computation of Test Cases from Integration-Effect Matrix

To compute the test cases from the integration-effect matrix, we have to calculate the number of test cases where an individual component is involved. It is calculated as:

Test cases for component C1, that is the case when C1 is involved

$$= (\text{Count of 1s below "Integration Effect" column in the first row}) - 1$$
$$= 4 - 1 = 3$$

Test cases for component C2, that is the case when C2 is involved

$$= (\text{Count of 1s below "Integration Effect" column in the second row}) - 1$$
$$= 4 - 1 = 3$$

TABLE 6.9

Probable Values of Integration-Effect Matrix of Five Components

Components	C1	Integration Effect	C2	Integration Effect	C3	Integration Effect	C4	Integration Effect	C5	Integration Effect
C1	1	Effect of (C1): 0/1	1	Integration effect of (C1 ∧ C2): 0/1	1	Integration effect of (C1 ∧ C3): 0/1	0	Integration effect of (C1 ∧ C4): 0/1	0	Integration effect of (C1∧ C5): 0/1
C2	0	Integration effect of (C2 ∧ C1): 0/1	1	Effect of (C2): 0/1	0	Integration effect of (C2 ∧ C3): 0/1	1	Integration effect of (C2 ∧ C4): 0/1	0	Integration effect of (C2 ∧ C5): 0/1
C3	0	Integration effect of (C3 ∧ C1): 0/1	0	Integration effect of (C3 ∧ C2): 0/1	1	Effect of (C3): 0/1	1	Integration effect of (C3 ∧ C4): 0/1	1	Integration effect of (C3 ∧ C5): 0/1
C4	1	Integration effect of (C4 ∧ C1): 0/1	0	Integration effect of (C4 ∧ C2): 0/1	0	Integration effect of (C4 ∧ C3): 0/1	1	Effect of (C4): 0/1	0	Integration effect of (C4 ∧ C5): 0/1
C5	0	Integration effect of (C5 ∧ C1): 0/1	0	Integration effect of (C5 ∧ C2): 0/1	0	Integration effect of (C5 ∧ C3): 0/1	0	Integration effect of (C5 ∧ C4): 0/1	1	Effect of (C5): 0/1

TABLE 6.10

Actual Values of Integration-Effect Matrix of Five Components

Components	C1	Integration Effect	C2	Integration Effect	C3	Integration Effect	C4	Integration Effect	C5	Integration Effect
C1	1	1	1	1	1	1	1	1	0	0
C2	1	1	1	1	1	1	1	1	0	0
C3	1	1	1	1	1	1	1	1	1	1
C4	1	1	1	1	1	1	1	1	1	1
C5	0	0	0	0	1	1	1	1	1	1

Test cases for component C3, that is the case when C3 is involved

$= ($Count of 1s below "Integration Effect" column in the third row$) - 1$
$= 5 - 1 = 4$

Test cases for component C4, that is the case when C4 is involved

$= ($Count of 1s below "Integration Effect" column in the fourth row$) - 1$
$= 5 - 1 = 4$

Test cases for component C5, that is the case when C5 is involved

$= ($Count of 1s below "Integration Effect" column in the fourth row$) - 1$
$= 3 - 1 = 2$

Overall test cases produced using integration-effect matrix:

= Count of test cases for first component C1 + Count of test cases for first
  component C2 + Count of test cases for first component C3
  + Count of test cases for first component C3 +
  Count of test cases for first component C5.
= 3 + 3 + 4 + 4 + 2 = 16.

When we apply the boundary-value analysis technique to this scenario we get:
If there are n number of components in the software, the least amount of test cases
produced by BVA are 4n + 1.
In the given scenario we have n = 5,
Then count of test cases = 4 * 4 + 1

= 4 * 5 + 1 = 21

6.5 Test-Case Generation Technique for White-Box Components

This testing is applicable to those components whose source code is available and accessible. To count the test-case number in conventional programs, McCabe's cyclomatic complexity formula is used, but in component-based software we do not have this technique. In this section we propose a method termed "cyclomatic complexity" to count the quantity of test cases for white-box components.

6.5.1 Cyclomatic Complexity for Component-Based Software

As discussed in Chapter 5, the cyclomatic complexity count is used to configure the number of independent paths in the program code, which in turn determines the test-case count needed to fulfill the testing of independent logics in the program code. There are two ways to define cyclomatic complexity in graph form.

- **Method 1:**
  Cyclomatic complexity, denoted V(G), where G is the flow graph, constructed according to the flow control of the components, can be defined as:

$$V(G) = |E| - |V| + 2 + |P|$$

where

$|E|$ denotes the number of edges, $|V|$ denotes the total number of vertices

$|P|$ is the total count of contributing components

Constant 2 is used to indicate that "the node V contributes to the complexity if its out-degree is 2"

- **Method 2:**

  In the second way of computing the cyclomatic complexity of component-based software, another metric can be used which is defined as:

  $$V(G) = \sum_{i=1}^{n} (IC)_i + \sum_{j=1}^{m} (CR)_j + OR$$

  where

  $(IC)_i = (IC_1, IC_2, IC_3, \ldots, IC_n)$ denotes cyclomatic complexities of all the contributing components,

  $(CR)_j = (CR_1, CR_2, CR_3, \ldots, CR_m)$ denotes the total closed regions found in the graph, and OR denotes the open region that is always 1.

6.5.2 Case Study

We need flow graphs to compute the cyclomatic complexity of component-based software. Notation for control-flow graphs and the base case of the proposed method are similar to those discussed in Chapter 5. We now describe a case study having six components: C1, C2, C3, C4, C5, and C6 (Figure 6.11). C1 is integrated with C2, C3, C4 and C6, forming closed regions CR1, CR4, CR5 and CR8, respectively. Component C2 has been integrated with C3 and that forms a closed region CR2. Component C3 has been linked with C5 and C6, making closed regions CR10 and CR15. Component C4 is associated with C5 and C6 and constructs closed regions CR9 and C13. Component C5 facing C6 is creating closed region CR12. The integration of components C1, C2 and C3 forms closed region CR3. Integration of C1, C3 and C5 forms closed region CR6. Integration of C1, C4 and C5 forms closed region CR7. Integration of C4 and C6 creates closed region CR11. Integration of components C3, C5 and C6 is constructing closed region CR14. There is an open region OR.

From Figure 6.10, it can be seen that the internal cyclomatic complexities of C1, C2, C3, C4, C5 and C6 are 4, 2, 4, 4, 2 and 2, respectively.

Count of vertices (V) = 44

Count of edges (E) = 70

Number of linked components (P) = 6

Now we compute the number of test cases in the case study defined in Figure 6.10, using McCabe's formula and the method discussed in this chapter, as:

- Number of test cases using McCabe method:

$$V(G) = e - n + 2p = 70 - 44 + 2*6 = 38$$

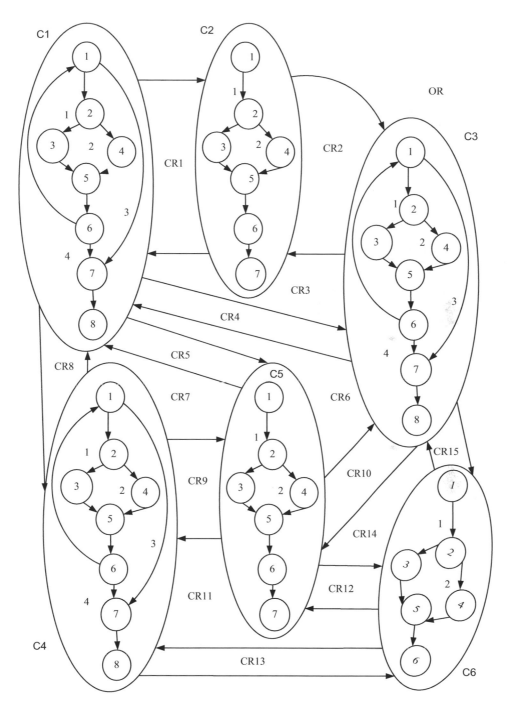

FIGURE 6.11
Interaction flow graph of six components.

Applying Method 1:
When the above-defined Method 1 is applied to the case study, we get:

$$V(G) = |E| - |V| + 2 + |P| = 70 - 44 + 2 + 6 = 34$$

Applying Method 2:
When the above-defined Method 2 is applied to the case study, we get:

$$n = 6, m = 15, IC_1 = 4, IC_2 = 2, IC_3 = 4, IC_4 = 4, IC_5 = 2, IC_6 = 2, CR = 1, \text{then}$$

$$V(G) = (4 + 2 + 4 + 4 + 2 + 2) + 15 + 1 = 34.$$

6.5.3 Performance Results

a. **Black-Box Testing:** When various case studies for different scenarios are analyzed in the context of black-box testing, it is observed that using the integration-effect matrix reduces the number of test cases compared to the boundary-value analysis (black-box) method, and better results are achieved. The results of both methods for varying numbers of components are shown in Table 6.11.

b. **White-Box Testing:** The performance of the cyclomatic complexity metric shown here is better than that of the methods defined in the literature. Compared with the McCabe method (white-box), better results are achieved in terms of reduced number of test cases. The results are shown in Table 6.12.

TABLE 6.11

Comparison of Black-Box Test Cases

Components	Boundary-Value Test Cases	Integration-Effect Metric Test Cases
2	9	2
3	13	4
4	17	10
5	21	16

TABLE 6.12

Comparison of White-Box Test Cases

Number of Components	McCabe's Complexity	Component-Based Cyclomatic Complexity
2 (1 closed region)	5	5
3 (2 closed regions)	12	11
3 (4 closed regions)	14	13
4 (3 closed regions)	15	13
4 (7 closed regions)	19	17
5 (10 closed regions)	27	24
6 (15 closed regions)	38	34

Summary

Whether it is traditional, object-oriented or component-based software, software testing is an essential software development process. Testing is performed not only to make the underlying software error free but to enhance and ensure the development of reliable and quality software products. In today's software development environment, testing commences just after the finalization of system requirements. Testing starts from component level and moves forward to the integrated complex system level. Testing techniques available in the literature are divided into two broad categories: black-box and white-box.

Testing is an attempt to make the software under consideration error free. Identifying and fixing errors in early phases of development is always helpful to minimize overall development costs. Testing software is expensive in terms of cost, effort and time. Testing is one of the critical and crucial phases of the overall development of software. Testing is the fundamental activity to verify the correctness, precision and compatibility of the software at individual level as well as system level. Practitioners have identified that improper testing results in untrustworthy and unreliable products.

Testing is commonly used to *verify* and *validate* software. *Verifying* components in CBSE constitutes the collection of procedures to certify the functionalities of components at individual level. *Validating* components is the group of procedures that ensures the integrity of integrated components according to the architectural design and fulfills the needs of the customer. Testing methods must incorporate test planning, test-case design, implementation of testing, results and the assessment of collected data. Typically, software is made up of different parts or sub-components. These subparts are not only independently deployed but are from various contexts. Therefore, we need testing techniques which address not only the subparts individually but the whole software.

This chapter includes two broad concepts for component-based software testing: testing methods and test-case generation techniques for component-based software.

Testing techniques discussed here are further categorized as white-box and black-box testing techniques. Testing levels for components and component-based software are introduced and the testing process for components and their levels is discussed in detail.

This chapter also includes two types of test-case generation techniques in the context of component-based software: the integration-effect matrix method for black-box testing and the cyclomatic complexity method for white-box testing are inter-module techniques applicable to interaction among various components. The integration-effect matrix is helpful for testing and recording the effects of components whose code is not accessible, and the cyclomatic complexity method is applicable to components for which code is available. There is a greater degree of predictability in terms of cost, effort, quality and risk if the testability of the software can be predicted properly and early.

References

Basili, V. R. 2001. "Cots-Based Systems Top 10 Lists." *IEEE Computer*, 34(5): 91–93.

Chen, J. 2011. "Complexity Metrics for Component-Based Software Systems." *International Journal of Digital Content Technology and Its Applications*, 5(3): 235–244.

Gill, N. S. and Balkishan. 2008. "Dependency and Interaction Oriented Complexity Metrics of Component Based Systems." *ACM SIGSOFT Software Engineering Notes*, 33(2): 1.

Gross, H.-G. 2009. *Component-Based Software Testing with UML*. Springer-Verlag, Berlin Heidelberg.

Harrold, M. J., D. Liang, and S. Sinha. 1999. "An Approach to Analyzing and Testing Component-Based Systems." In *Workshop on Testing Distributed Component Based Systems (ICSE 1999)*, Los Angeles, CA.

Myers, G. 1979. *The Art of Software Testing*. Wiley, Hoboken, NJ.

Umesh, K.T. and S. Kumar. 2017, "Component Level Testing in Component-Based Software Development." *International Journal of Innovations & Advancement in Computer Science*, 6(1): 69–73.

Weyuker, E. J. 1998. "Testing Component-Based Software: A Cautionary Tale." *IEEE Software*, 15(5): 54–59.

Weyuker, E. J. 2001. "The Trouble with Testing Components." In Heineman, G. T. and Councill, W. T., eds., *Component-Based Software Engineering*. Addison-Wesley, Reading, MA.

7

Estimating Execution Time and Reliability of Component-Based Software

7.1 Introduction

Component-based software development is a proficient archetype for constructing quality software products. Reusability is the basic concept behind the component-based software environment where components interact to share services and information. Reliability of commercial off-the-shelf (COTS) component-based software systems may depend on the reliabilities of the individual commercial off-the-shelf (COTS) components that are in the system (Hamlet et al. 2001). This chapter describes a reliability estimation method that uses reusability features of components on an execution path or execution history. Function points are used as the basic metric to estimate the reusability metric of the different categories of components. In the absence of function points, we can use lines of code to define the reusability metric.

In this chapter, the focus is on the interaction aspect of components, which can be shown by an interaction graph. An interaction metric is a measurement of the actual execution time of individual components in component-based software. Interaction metrics are easy to compute, and are informative when analyzing the performance of components as well as component-based software.

7.2 Interaction Metric

In component-based software, components communicate with each other. The manner of their interaction depends upon various issues, including the pre-specified architectural path design, implementation set of a particular application, operational profiles of users and the set of defined inputs (D'Souza and Wills 1998, Michael 2000, Philippe 2003, Hopkins 2000). There is more than one possible path in component-based software applications, and a particular path is selected from a number of probable paths. Execution of a particular path requires the invocation of components residing on that path, from source component to destination component. Components call or are called by each other either sequentially or using backward and forward loops (Meyer 2003). These invocations and interactions must be taken into account when estimating the execution time of the overall software.

In the literature, the importance of interactions among components and software constructs has been broadly explained in terms of cohesion and coupling, information flow, cyclomatic complexity, function-point analysis and many other aspects (Vitharana 2003, Basili and Boehm 2001, Jianguo et al. 2001, Tyagi and Sharma 2012). Distinguished researchers have suggested various categories of metrics to compute the interaction complexities of software. Chidamber and Kemerer proposed a set of metrics to count the number of instant derived classes, the longest path from start node, total methods residing in a class, number of coupled classes, the ratio of difference in methods in a class and the number of methods executing in response after invoking a class (Jacobson et al. 1997, McIlroy 1976). Misra (1992) defined a metric to estimate the method-level complexity of a class by weighing up the internal structure of the method. Fothi et al. (2003) proposed a metric to analyze class complexity taking control structures, data and the relationship between them into account. For component-based software, practitioners have proposed various estimation metrics considering the interaction, coupling and cohesion properties of components (Gill and Balkishan 2008, Noel 2006, Sengupta and Kanjilal 2011). Gill and Balkishan (2008) suggested a collection of component-based metrics to assess the interaction dependency and coupling properties of components. Their work includes metrics for component dependency and density of component interaction.

7.2.1 Basic Definitions and Terminologies

The terms defined here are used to develop execution time and reliability estimation metrics. The following abbreviations are used:

Abbreviation	Term
C_i	Component
Re-C_i	Reliability of a component
RMC_i	Reusability metric of a component
$InvC_i$	Invocation counts of the component
ET-C_i	Execution time of component
EP_i	Execution of a path
ET-P_i	Execution time of a path
ET-CBS	Execution time of component-based software
EP-P_i	Probability of an execution of a path
$ReqI_{i,j}$	Request interaction
$ResI_{i,j}$	Response interaction
IM-$C_{i,j}$	Interaction metric between two components
PI-$C_{i,j}$	Probability of interaction between two components
Av-ET-C_i	Average execution time of a component
Ac-ET-P_i	Actual execution time of path
Ac-ET-CBS	Actual execution time of CBS
$RMC_{i\text{-Full-qualified}}$	Fully qualified reusability metric
$RMC_{i\text{-Part-qualified}}$	Partially qualified adaptable reusability metric
$RMC_{i\text{-Off-the-shelf}}$	Off-the-shelf reusability metric
Re-P_i	Reliability of a path
Re-CBS	Reliability of component-based software

Terms listed above in the context of execution time and reliability estimations can be defined as follows:

a. **Node or Component (C_i):** A component is the fundamental artifact of component-based software.

b. **Reliability of a Component (Re-C_i):** Reliability of a Component (Re-C_i) represents the probability of failure-free execution of component C_i, where i = 1, 2, 3, ..., n. In this work, we assume that the reliability of components is known.

c. **Reusability Metric of a Component (RMC_i):** The reusability metric of component C_i is shown as RRC_i, where i = 1, 2, 3..., n. The reusability metric of a component depends on its level of reusability. It may be off-the-shelf, fully qualified, partially qualified or new, as discussed in Chapter 4.

d. **Invocation counts of the component ($InvC_i$):** In component-based software, components invoke or call each other to access and provide functionalities. A component can invoke itself through a loop or recursive call. $InvC_i$ is the number of times a component has been executed or invoked, either to access or to provide information and services.

e. **Execution Time of Component (ET-C_i):** Every component takes a length of time to be executed. ET-C_i denotes the time taken by the component to perform its desired and defined services.

f. **Execution of a Path (EP_i):** Path execution is the sequence of execution of components in a particular path from source to termination. Execution of a particular path represents the way components interact with each other. If the path has k components from start to end and we have n paths, execution of a path is defined as:

$$EP_1 = C_{11} \rightarrow C_{12} \rightarrow C_{13} \rightarrow \cdots \rightarrow C_{1k}$$

$$EP_2 = C_{21} \rightarrow C_{22} \rightarrow C_{23} \rightarrow \cdots \rightarrow C_{2k}$$

$$EP_3 = C_{31} \rightarrow C_{32} \rightarrow C_{33} \rightarrow \cdots \rightarrow C_{3k}$$

$$\cdots$$

$$EP_n = C_{n1} \rightarrow C_{n2} \rightarrow C_{n3} \rightarrow \cdots \rightarrow C_{nk}$$

where, k = 1, 2, 3, ..., t. t is the termination node. Here, it is important to note that the number of components in different paths may vary. Therefore, the execution time of each execution history is different.

g. **Execution Time of a Path (ET-P_i):** Execution time of an execution history is the sum of the execution times of all its participating components. We calculate the execution time of a path as:

$$ET - P_i = \sum_{i=1}^{|t|} \left(ET - C_i * InvC_i \right) \tag{7.1}$$

where (ET-C_i) is the execution time of a component and $InvC_i$ is the number of invocations. Since the execution history may contain backward and forward loops, we have taken the number of invocations of a component into account. The minimum value of $InvC_i$ is 1, since a component lying on a particular path must be invoked at least once.

h. **Execution Time of Component-based software (ET-CBS):** Execution time of component-based software is defined as the time taken by the software to execute all its paths. It is denoted ET-CBS, where CBS denotes the component-based software and E_T is the execution time.

i. **Probability of Execution of a Path (EP-P_j):** There is more than one path possibility in CBS applications. The selection probability of a particular execution history is considered from a number of probable execution histories. Execution of an execution history requires the invocation of components residing on that path, from start to end, one after the other, or using backward and forward loops (Littlewood and Strigini 1993).

j. **Request Interaction (ReqI$_{i,j}$):** Request interactions are shown as directed edges going outwards to the component C_i. They represent the request of services by the component or are invocations made by the component C_i to some other component C_j. They are denoted ReqI$_{i,j}$. They contain the parameters required to make the request (Figure 7.1).

$$ReqI_{i,j} = \left| \text{Total number of request parameters going from} \, C_i \, \text{to} \, C_j \right|$$

k. **Response Interaction (ResI$_{i,j}$):** Response interactions are the directed edges incoming to the component C_i. They show the response to requests or invocations made by component C_j to component C_i. They are shown as ResI$_{i,j}$. They contain the total number of response parameters required to fulfill the response (Figure 7.1).

$$ResI_{i,j} = \left| \text{Total number of response parameters coming from} \, C_j \, \text{to} \, C_i \right|$$

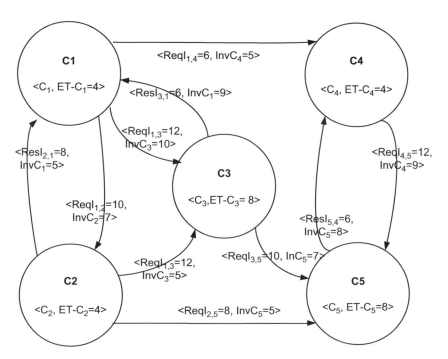

FIGURE 7.1
Interactions among five components.

l. **Interaction Metric between two Components, C_i and C_j (IM-$C_{i,j}$):** The interaction metric between two components can be calculated by taking the ratio between the total number of "res-interactions" and the sum of "req-interactions" and "res-interactions" of components.

$$IMC_{i,j} = \frac{|ResI_{i,j}|}{|ReqI_{i,j}| + |ResI_{i,j}|}$$

m. **Probability of Interaction between two components (PI-$C_{i,j}$):** Suppose there are two components, C_i and C_j. The interaction probability is the probability of selection of component C_j for execution, after component C_i has been executed. In other words, the interaction probability is the probability that component C_i invoked component C_j to perform its services. The sum of total probability of interactions invoked by a component to other components is always 1. A predefined execution history predetermines the probability of interaction among components as unity.

7.2.2 Average Execution Time of a Component

Components are invoked to access or to provide services. On a path P, a component can be invoked as many times as its services are required by the software system. Average execution time of a component is computed as

$$Av - ET - C_i = ET - C_i \times InvC_i \qquad (7.2)$$

where ET-C_i is the execution time of component C_i and $InvC_i$ is the count of invocations of component C_i.

7.2.3 Execution Time of a Path

The execution time of a path is the sum of average execution times of all its participating components. We denote execution time of path Pi as ET-Pi. We calculate the execution time of a path as

$$ET - P_i = \sum_{i=1}^{n} \left(ET - C_i * InvC_i \right) \qquad (7.3)$$

Since the path may contain backward and forward loops, we have taken the number of invocations of a component into account. The minimum value of $InvC_i$ is 1, since a component lying on a particular path must be invoked at least once.

7.2.4 Actual Execution Time of Paths

It is important to note that during the execution of a path P, components interact with each other through request and response parameters. Their interaction metrics therefore play an important role in calculating the execution time. We have taken the interaction metric as an additional factor which includes the time and overhead of the interaction between two components. Every time a component C_i interacts with another component C_j, their interaction metric IM-$C_{i,j}$ is computed. Now we define the actual execution time of a path P_i as

$$Ac-ET-P_i = \sum_{i=1}^{n} \left(\left(E-C_i * InvC_i \right) + \left(ET-C_j * InvC_j \right) + IM-C_{ij} \right) \tag{7.4}$$

where Ac represents the actual time. C_i and C_j are two consecutive components lying on the same path P_i. Component C_i invokes component C_j and C_j responds. There is an interaction between C_i and C_j so their interaction metric IM-$C_{i,j}$ is also calculated. ET-C_i is the execution time of the component; InvC$_i$ is the invocation count. Without the loss of generality, we assume that the default value of the interaction metric between two components IM-$C_{i,j}$ is 0.01, if interaction is one way.

7.2.5 Actual Execution Time of CBS

In a component-based software application, we may have more than one possible path. Actual execution time for component-based software is defined as the sum of the execution of all the paths, including their interaction metric, that is

$$Ac-ET-CBS = \sum_{P_i=1}^{m} \left\{ Ac-ET-P_i \right\} \tag{7.5}$$

where i = 1 to n, that is the number of components on path P_i, InvC$_i$ is the number of invocations of component i and IM-$C_{i,j}$ is the interaction metric computed between components C_i and C_j.

7.3 Case Study and Discussion

To illustrate the proposed method, we use a case study which is defined in Figure 7.1. This case study has five components. Every component has some basic information regarding its execution time, number of invocations, number of request and response interactions. Execution time of component C1 is 4, C2 is 4, C3 is 8, C4 is 4 and C5 is 8 units of time. Components are invoked by each other through request and response interactions. Lying on different paths, components have different invocation counts.

A component can be invoked by other components through request interactions and component responses through response interactions. It is possible for a component to have only outgoing edges and no incoming edges. In such cases we assume that the interaction-metric value between these components is 0.01. An interaction request between two components i and j is shown as $ReqI_{i,j}$ and the response interaction is shown as $ResI_{j,i}$. The invocation count among components is represented as Inv. Every interaction contains two-tuple information: the type of interaction, that is, request or response; and the number of invocations made by that component. These interactions contain the number of parameters used to request or respond.

Figure 7.1 shows that there are four execution paths from the starting component C_1 to the ending component C_5.

These paths are

$P_1 \rightarrow C_1 \rightarrow C_2 \rightarrow C_5$
$P_2 \rightarrow C_1 \rightarrow C_3 \rightarrow C_5$
$P_3 \rightarrow C_1 \rightarrow C_2 \rightarrow C_3 \rightarrow C_5$
$P_4 \rightarrow C_1 \rightarrow C_4 \rightarrow C_5$

- **Values and computations of path P1: P1→C1→C2→C5**

  Path P_1 contains three components, C_1, C_2 and C_5. Table 7.1 contains the values for execution time, number of invocations and the average execution times of the components lying on the path P_1. Computations of requests from one component to other components, responses from called components and the calculation of interaction metrics are given in Table 7.2.

- **Computations of requests, responses and interaction metrics of path P1**

  Applying the values of Tables 7.1 and 7.2 to Equation (7.4), we calculate the actual execution time of path P_1 as

$$Ac-ET-P_1 = \left(\left(ET-C_1 * InvC_1\right) + \left(ET-C_2 * InvC_2\right) + IM-C_{1,2} \right)$$
$$+ \left(\left(ET-C_2 * InvC_2\right) + \left(ET-C_5 * InvC_5\right) + IM-C_{2,5} \right)$$
$$= \left(\left(4*5\right) + \left(4*7\right) + 0.4\right) + \left(\left(4*7\right) + \left(8*5\right) + 0.01\right)$$

$$Ac-ET-P_1 = 48.4 + 68.01 = 116.41$$

- **Values and computations of path P2: P2→C1→C3→C5**

  The second path, P_2, also contains three components, C_1, C_3 and C_5. Values for execution time, number of invocations and the average execution times of the components

TABLE 7.1

Inputs of Path P_1

C_i	ET-C_i	InvC$_i$	Av-ET-C_i
C_1	4	5	20
C_2	4	7	28
C_5	8	5	40

TABLE 7.2

Interaction Metrics of Path P_1

$C_i \rightarrow C_j$	ReqI$_{i,j}$	ResI$_{j,i}$	IM-$C_{i,j}$
$C_1 \rightarrow C_2$	10	8	0.4
$C_3 \rightarrow C_5$	8	0	0.01

TABLE 7.3

Inputs of Path P_2

C_i	ET-C_i	InvC_i	Av-ET-C_i
C_1	4	9	36
C_3	8	10	80
C_5	8	7	56

TABLE 7.4

Interaction Metrics of Path P_2

$C_i \rightarrow C_j$	ReqI$_{i,j}$	ResI$_{j,i}$	IM-$C_{i,j}$
$C_1 \rightarrow C_3$	12	6	0.3
$C_3 \rightarrow C_5$	10	0	0.01

lying on path P_2 are given in Table 7.3. Calculations of requests from one component to other components, responses from called components and the interaction metrics are given in Table 7.4.

- **Computations of requests, responses and interaction metrics of path P2**

  Values defined in Tables 7.3 and 7.4 are applied to Equation (7.4) to calculate the actual execution time of path P_2, as

$$\begin{aligned} Ac-ET-P_2 &= \Big(\big(ET-C_1 * InvC_1\big)+\big(ET-C_3 * InvC_3\big)+IM-C_{1,3}\Big) \\ &\quad +\big(ET-C_3 * InvC_3\big)+\big(ET-C_5 * InvC_5\big)+IM-C_{3,5}\big) \\ &= \Big(\big(4*9\big)+\big(8*10\big)+0.3\Big)+\Big(\big(8*10\big)+\big(8*7\big)+0.01\Big) \end{aligned}$$

$$Ac-ET-P_2 = 116.3+136.01 = 252.31$$

- **Values and computations of path P3: P3→C1→C2→C3→C5**

  The third path, P_3, has four components, C_1, C_2, C_3 and C_5. Values for execution time, number of invocations and the average execution times of the components lying on path P_3 are given in Table 7.5. Requests and responses from components and the computation of interaction metrics are given in Table 7.6.

- **Computations of requests, responses and interaction metrics of path P3**

  When the values of Tables 7.5 and 7.6 are applied to Equation (7.4), the actual execution time of path P_3 is calculated as

$$\begin{aligned} Ac-ET-P_3 &= \Big(\big(ET-C_1 * InvC_1\big)+\big(ET-C_2 * InvC_2\big)+IM-C_{1,2}\Big) \\ &\quad +\Big(\big(ET-C_2 * InvC_2\big)+\big(ET-C_3 * InvC_3\big)+IM-C_{2,3}\Big) \\ &\quad +\Big(\big(ET-C_3 * InvC_3\big)+\big(ET-C_5 * InvC_5\big)+IM-C_{3,5}\Big) \\ &= \Big(\big(4*5\big)+\big(4*7\big)+0.4\Big)+\Big(\big(4*7\big)+\big(8*5\big)+0.01\Big)+\Big(\big(8*5\big)+\big(8*5\big)+0.01\Big) \end{aligned}$$

$$Ac-ET-P_3 = 48.4+68.01+80.01 = 196.42$$

TABLE 7.5

Inputs of Path P_3

C_i	ET-C_i	InvC_i	Av-ET-C_i
C_1	4	5	20
C_2	4	7	28
C_3	8	5	40
C_5	8	5	40

TABLE 7.6

Interaction Metrics of Path P_3

$C_i \rightarrow C_j$	ReqI$_{i,j}$	ResI$_{j,i}$	IM-$C_{i,j}$
$C_1 \rightarrow C_2$	10	8	0.4
$C_2 \rightarrow C_3$	12	0	0.01
$C_3 \rightarrow C_5$	10	0	0.01

- **Values and computations of path P4: P4→C1→C4→C5**
  The fourth path, P_4, has three components, C_1, C_4 and C_5. The values for execution time, number of invocations and the average execution times of the components lying on path P_4 are given in Table 7.7. Calculations of requests from one component to other components, responses from called components and the interaction metric are given in Table 7.8.

- **Computations of requests, responses and interaction metrics of path P4**

  When the values of Tables 7.7 and 7.8 are applied to Equation (7.4), the actual execution time of path P_4 is computed as

$$Ac-ET-P_4 = \left(\left(ET-C_1 * InvC_1 \right) + \left(ET-C_4 * InvC_4 \right) + IM-C_{1,4} \right)$$
$$+ \left(\left(ET-C_4 * InvC_4 \right) + \left(ET-C_5 * InvC_5 \right) + IM-C_{4,5} \right)$$

TABLE 7.7

Inputs of Path P_4

C_i	ET-C_i	InvC_i	Av-ET-C_i
C_1	4	1	4
C_4	4	5	80
C_5	8	9	56

TABLE 7.8

Interaction Metrics of Path P_4

$C_i \rightarrow C_j$	ReqI$_{i,j}$	ResI$_{j,i}$	IM-$C_{i,j}$
$C_1 \rightarrow C_4$	6	0	0.01
$C_4 \rightarrow C_5$	12	6	0.3

TABLE 7.9

Actual Execution Time of Four
Different Paths

Path (P_i)	Ac-ET-P_i
P_1	116.41
P_2	252.31
P_3	196.42
P_5	116.31

$$= \left((4*1) + (4*5) + 0.01 \right) + \left((4*5) + (8*9) + 0.7 \right)$$

$$Ac - ET - P_4 = 24.01 + 92.3 = 116.31$$

- **Actual execution time of component-based application**

  The actual execution time of component-based software given in Figure 7.2 is computed by applying the values of Table 7.9 to Equation (7.5), as

  $$Ac - ET - CBS = Ac - ET - P_1 + Ac - ET - P_2 + Ac - ET - P_3 + Ac - ET - P_4$$

  $$= 681.45$$

7.4 Reliability Estimation

In software engineering, reliability is the foundation stone of the software's quality. Reliability can be defined as:

> the probability of failure-free software operation in a specified environment for a specified period of time.

(Lyu 1996)

Another definition of reliability is:

> the probability of failure-free operation of a computer program for a specified time in a specified environment.

(Henry and Kafura 1981)

Musa defined reliability as:

> the likelihood of execution without failure for some definite interval of natural units or time.

(Musa 1998)

According to IEEE:

> Software reliability is defined as the ability of a system or a component to perform its required functions under stated conditions for a specified period of time.

(IEEE Standard 610.12-1990 1990)

Software reliability and reliability issues are discussed in detail in Chapter 3. Software reliability estimation techniques are divided into two broad categories: time dependent and time independent.

a. **Time-dependent reliability:**
   These techniques assume that reliability is dependent on time. As time passes, reliability of "things" decreases.
   Among the reliability growth models most used in academia and industry are the exponential model and the logarithmic Poisson model.
   - *Exponential model*: Exponential models rely on the concept that finite failures can occur in infinite time.
   - *Logarithmic Poisson model*: These models are based on the assumption that infinite failures can occur in infinite time.

b. **Time-independent reliability:**
   This type of reliability model assumes that reliability does not depend on time, especially in the case of software reliability, but on how software behaves on certain inputs.

7.4.1 Hardware Reliability and Software Reliability

In practice, software reliability is different from hardware reliability. Hardware reliability depends upon time and various environmental factors. In the early stages of hardware, error rates and failure rates are high, in some cases due to design or manufacturing faults or defects. The life cycle of hardware products can be divided into three phases: burn-in phase, useful life phase and wear-out phase (Figure 7.2).

- **Burn-in Phase:** This is the phase when hardware products try to cope with environmental factors. In the earlier stages, errors and failure rates are at their peak. With time and changes in the product as well as changes in the environment, the failure rate becomes constant. Hardware products then enter the useful life phase.

- **Useful Life Phase:** Once initial failures and errors are fixed, the useful life phase of the product commences, when it can be used without major faults. Minor errors may be encountered and fixed without major effort.

- **Wear-Out Phase:** After a certain period of time, due to environmental factors like dust, regression, temperature, weather, lack of maintenance and similar, the product may encounter major failures and the need for replacement arises. Hardware products can become outdated due to environmental maladies, that is, they start to wear out.

In software, we expect to have ideal curves, as software is not susceptible to wear-out due to environmental maladies. Once the software comes into use, errors may be generated due to poor design or coding bugs. But after all errors have been fixed, software exhibits normal behavior and can run for an infinite time. This is the expected curve for software, referred to as an ideal curve (Figure 7.3).

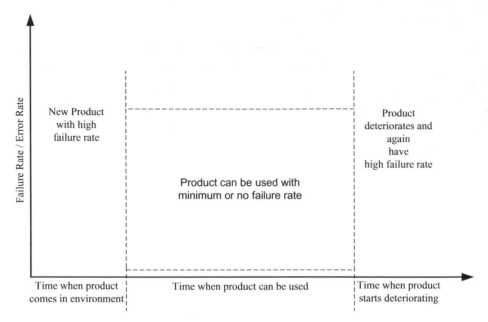

FIGURE 7.2
Hardware reliability. (Adapted from Klutke et al. 2003.)

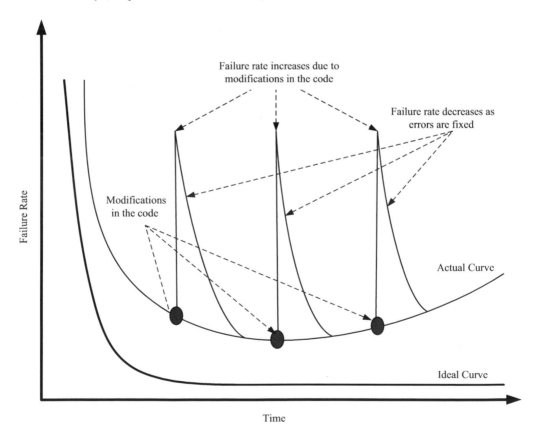

FIGURE 7.3
Software reliability.

But in practice, as time passes, user requirements and demands change, and developers attempt to modify the software to fulfill them. As the changes are made, new faults and errors may be introduced and the failure rate rises again. Errors are removed and the failure rate again settles at constant level. After a period of time, new requirements may introduce new errors. Again, these are fixed and make the software reusable. In this way software is said to deteriorate rather than to wear out. These repeated activities continue and maintenance activities keep software usable. This is called an actual curve (Figure 7.3).

The major differences between hardware and software reliabilities are:

- Generally, hardware reliability is a function that depends on time, whereas software reliability is not time dependent.
- Hardware wears out due to environmental maladies, whereas software deteriorates due to changes in requirements or similar modifications in the software code.
- Hardware components are replaced by new components when their reliability decreases and failure rates increase, whereas no such concept exists in software. Software components are extended rather than replaced.
- Every hardware/hardware product suffers from aging factors, that is, after a certain time it must be replaced, whereas in software there is no aging factor. Software can work for an infinite time where there is no change in the requirements.
- Measuring hardware reliability is easy as there are physical factors, parameters and metrics, whereas software reliability assessment is quite difficult. Hardware-oriented measures and metrics are not applicable to software products.

7.4.2 Reliability Measures and Metrics Available

As we have seen, reliability estimations of software are quite different from software reliability assessment. In the context of hardware we have direct assessment measures, whereas in software, reliability is an indirect attribute. Also, there is no single universally accepted software reliability assessment metric for all types of software. In this section we discuss well-known software reliability estimation metrics.

a. **Rate of Occurrence of Failure (ROCOF):** ROCOF is the number of failures that occur in software in a specified period of time. For reliable software there must be as few failures as possible. As the failure count increases, reliability of the software system decreases.

b. **Mean Time to Failure (MTTF):** MTTF is the average time between failure occurrences in operational software. Software running for a long time may produce a number of failures. MTTF is the metric used to keep track of the average time between successive failures. As the MTTF value increases, so does the reliability of the software.

c. **Mean Time to Repair (MTTR):** MTTR is the average time taken to repair the failure and make the software operational. As the repair time increases, the loss due to failure also increases. Therefore for better reliability of the software system, the MTTR should be less.

d. **Mean Time between Failures (MTBF):** This is defined as the sum of MTTF and MTTR. That is, MTBF = MTTF + MTTR.

e. **Probability of Failure on Demand (POFOD):** This is defined as the probability of failure of software when a request for service occurs. This is different from other metrics as it does not involve time. If the software is designed in a way that is free from service requests, then such a metric is of no use. But if the service requests are quite frequent, then this metric is quite useful to predict the probability of on-demand failure.

f. **Software Availability:** This is the metric used to assess the time for which the system is likely to be available for operations over a period of time. It is defined as the ratio of MTBF and sum of MTBF and MTTR. That is:

$$\text{Software availability} = \frac{\text{MTBF}}{\text{MTBF} + \text{MTTR}}$$

7.5 Reliability Estimation Method Using Reusability Metrics

In this section, a reusability metric is introduced as an aspect of reliability prediction. This book has emphasized the reusability feature of components and component-based software, using the reusability metric for new, adaptable (both fully qualified and partially qualified) and off-the-shelf components. The interaction metric (Section 7.2.1) is used as the second aspect of the reliability assessment of component-based software. We introduce the role of interaction metric for individual components to estimate the execution time of component-based software. On the basis of the structure of interactions between components used in development, a component interaction graph is constructed. The formation of the graph is in accordance with the probability of interaction of components as well as the probability of the selection of different paths available in the component interaction graph.

The reliability estimation discussed in this section is based on the following factors:

i. Reusability metric of components

ii. Reliabilities of individual components

iii. Actual execution time of an execution history

iv. The probability of selection of a particular path's execution history

v. The execution time of each component on the selected execution history

vi. Actual execution time of the CBS application

vii. The invocation counts of the components, considering that the components are called by other components to access and provide services

viii. Reliability of execution history

ix. Reliability of CBS application

7.5.1 Component Reusability Metric

As we have seen, component-based software is a mixture of new and reused components. The aim of this reliability estimation technique is to analyze and apply the reusability feature of components and component-based software. The reusability metric is the trade-off between new and reused components. The reusability metric can be defined using function points or lines of code. This section discusses only those reusability metrics which are applicable to reliability estimation. For a detailed discussion of other reusability metrics, see Chapter 4.

7.5.1.1 *Fully Qualified Reusability Metric at Component Level* ($RMC_{i\text{-}Full\text{-}qualified}$)

SCENARIO 1: COMPONENT-LEVEL ADAPTABLE REUSABILITY METRIC ($RMC_{i\text{-}FULL\text{-}QUALIFIED}$)

Fully qualified components require little alteration to fit into new contexts. At component level, reusability is assessed in terms of function points. There are three cases of reusability metric for fully qualified components:

Case 1: Reusability metric when all parts of the component are involved
Case 2: Reusability metric when only reused parts of the component are involved
Case 3: Reusability metric when only adaptable parts of the component are involved

Each case can be defined using function points as well as lines of code.

i. **When all parts of the component are involved**
Using function points:

In this context, the ratio of fully qualified function points of a particular component C_i to the total number of function points of that component is taken. This is defined as:

$$RMC_{i\text{-}Full\text{-}qualified} = \frac{\left|FPC_{i\text{-}Full\text{-}qualified}\right|}{\left|FPC_i\right|}$$

where FPC_i denotes the total function points of a component including new, adaptable and reused.

Using lines of code (LOC):

On the basis of lines of code (LOC), we can define the metric as

$$RMC_{i\text{-}Full\text{-}qualified} = \frac{\left|LOCC_{i\text{-}Full\text{-}qualified}\right|}{\left|LOCC_i\right|}$$

where $LOCC_i$ denotes the total LOC of a component including new, customized and reused.

ii. **When only reused parts of the component are involved**
Using function points:

In this context the ratio is taken with respect only to the reused function points of the component. Here the total number of fully qualified function points is divided by the total number of reused function points.
Therefore, the reusability metric in terms of function points is defined as:

$$RMC_{i\text{-}Full\text{-}qualified} = \frac{\left|FPC_{i\text{-}Full\text{-}qualified}\right|}{\left|FPC_{i\text{-}Reused}\right|}$$

where total reused function points $FPC_{i\text{-}Reused}$ represent the collection of off-the-shelf, fully qualified and partially qualified adaptable components.

Using lines of code (LOC):

On the basis of lines of code (LOC), we can define the metric as

$$RMC_{i-Full-qualified} = \frac{\left|LOCC_{i-Full-qualified}\right|}{\left|LOCC_{i-Reused}\right|}$$

where $LOCC_{i-Reused}$ denotes the collection of off-the-shelf, fully qualified and partially qualified reusable LOCs.

iii. **When only adaptable parts of the component are involved**
Using function points:

In this case the ratio is taken in the context of adaptable function points only. The total number of fully qualified function points is divided by the total number of adaptable function points of the component. Therefore, the reusability metric is defined as

$$RMC_{i-Full-qualified} = \frac{\left|FPC_{i-Full-qualified}\right|}{\left|FPC_{i-Adaptable}\right|}$$

where adaptable function points $FPC_{i-Adaptable}$ are the assembly of fully qualified and partially qualified function points only.

Using lines of code :

On the basis of lines of code (LOC), we can define the metric as

$$RMC_{i-Full-qualified} = \frac{\left|LOCC_{i-Full-qualified}\right|}{\left|LOCC_{i-Adaptable}\right|}$$

where adaptable LOCs are the assembly of fully qualified and partially qualified LOCs only.

7.5.1.2 *Partially Qualified Adaptable Reusability Metric at Component Level (RMC$_{i\text{-}Part\text{-}qualified}$)*

For partially qualified components, two scenarios are discussed below.

SCENARIO 1:   COMPONENT-LEVEL ADAPTABLE REUSABILITY METRIC (RMC$_{I\text{-}PART\text{-}QUALIFIED}$)

We define the reusability metric for partially qualified adaptable components at component level in three different contexts:

Case 1: Reusability metric when all parts of the component are involved
Case 2: Reusability metric when only reused parts of the component are involved
Case 3: Reusability metric when only adaptable parts of the component are involved

Each case can be defined using function points as well as lines of code.

i. **When all parts of the component are involved**
Using function points:

In this case, we calculate the reusability metric in the context of a particular component when all parts of the components are involved. Here, the total number of partially qualified reused function points is divided by the total number of function points of the components. That is:

$$RMC_{i-Part-qualified} = \frac{\left|FPC_{i-Part-qualified}\right|}{\left|FPC_i\right|}$$

Using lines of code:

Here, the total number of partially qualified reused lines of code is divided by the total number of lines of code of the components. That is:

$$RMC_{i-Part-qualified} = \frac{\left|LOCC_{i-Part-qualified}\right|}{\left|LOCC_i\right|}$$

ii. **When only reused parts of the component are involved**
Using function points:

In this context the ratio is taken with respect to reused function points of the component only. In this case the total number of partially qualified function points is divided by the total number of reused function points. Hence, it is defined as

$$RMC_{i-Part-qualified} = \frac{\left|FPC_{i-Part-qualified}\right|}{\left|FPC_{i-Reused}\right|}$$

Using lines of code (LOC):

In this case the total number of partially qualified lines of code is divided by the total number of reused lines of code. Hence, it is defined as

$$RMC_{i-Part-qualified} = \frac{\left|LOCC_{i-Part-qualified}\right|}{\left|LOCC_{i-Reused}\right|}$$

iii. **When only adaptable parts of the component are involved**
Using function points:

In this case, the ratio is taken in the context of adaptable function points only. The total number of fully qualified function points is divided by the total number of adaptable function points only. Therefore, the reusability metric is defined as

$$RMC_{i-Part-qualified} = \frac{\left|FPC_{i-Part-qualified}\right|}{\left|FPC_{i-Adaptable}\right|}$$

Using lines of code:

The total number of fully qualified lines of code is divided by the total number of adaptable lines of code only. Therefore, the reusability metric is defined as

$$RMC_{i-Part-qualified} = \frac{|LOCC_{i-Part-qualified}|}{|LOCC_{i-Adaptable}|}$$

7.5.1.3 Off-the-Shelf Reusability Metric at Component Level (RMC_{i-Off-the-shelf})

Off-the-shelf components are those that can be reused without any modification.

Component-level off-the-shelf reusability metric (RMC_{i-Off-the-shelf})

At component level there are two different contexts for reusability metrics: when all parts of the component are involved, and when only reused parts are involved. The third context, when only adaptable components are involved, is not applicable here.

Case 1: Reusability metric when all parts of the component are involved

Case 2: Reusability metric when only reused parts of the component are involved

Each case can be defined using function points as well as lines of code.

i. **When all parts of the component are involved**
   Using function points

   In this context, we compute the reusability metric for a particular component. The total number of off-the-shelf reused function points is divided by the total function points of the component. It is defined as:

   $$RMC_{i-Off-the-shelf} = \frac{|FPC_{i-Off-the-shelf}|}{|FPC_i|}$$

   Using lines of code

   The total number of off-the-shelf reused lines of code is divided by the total number of lines of code of the component. It is defined as

   $$RMC_{i-Off-the-shelf} = \frac{|LOCC_{i-Off-the-shelf}|}{|LOCC_i|}$$

ii. **When only reused parts of the component are involved**
    Using function points:

    This is the case when the ratio is taken in the context of only reused function points of the component. Here, the total number of off-the-shelf function points is divided by the total reused function points. It is given as

    $$RMC_{i-Off-the-shelf} = \frac{|FPC_{i-Off-the-shelf}|}{|FPC_{i-Reused}|}$$

Using lines of code:

The total number of off-the-shelf lines of code is divided by the total number of reused lines of code. It is given as

$$RMC_{i-Off-the-shelf} = \frac{\left|LOCC_{i-Off-the-shelf}\right|}{\left|LOCC_{i-Reused}\right|}$$

7.5.2 Reliabilities of Individual Components

In this reliability estimation technique, it is assumed is that the reliability of components is known, and the reliability of each component is denoted Re-C_i.

7.5.3 Reliability of a Path (Re-P_i)

On a particular path, we have more than one component, and every component has its reliability. Components cooperate with each other; hence there is some degree of interaction among them. In addition, when components interact more than once, their reusability is also an important issue regarding their reliability. During the computation of reliability of an execution history, we introduce these two factors, that is, the interaction metric and the reusability metric. The reliability of an execution history of a particular path is calculated as

$$Re-P_i = \sum_{j=1}^{|t|} \prod_{i=1}^{j} \left\{ PI-C_{ij} * \left(Re-C_i + \left(1 - Re-C_i^{RMC_i * IM-C_i} \right) \right) \right\} \tag{7.6}$$

where PI-C_{ij} is the probability of invocation of component C_i, Re-C_i is the reliability of component C_i, RMC$_i$ is the reusability metric of component C_i and IM-C_{ij} is the interaction metric of components i, which may be new, adaptable or off the shelf. Without the loss of generality, we assume that the reusability metric of new components is 0.01.

7.5.4 Reliability of Component-Based Software (Re-CBS)

It can be concluded that the reliability of component-based software is a function of the reliabilities of individual components (Lyu 1996, Malaiya et al. 2002, Misra 1992).

When there is more than one execution history in a CBS application, each history has its own reliability estimated from the starting component to the ending component. We define the reliability of component-based software as the probability of each path and the reliability of that path's execution history, that is:

$$Re-CBS = \sum_{Hi=1}^{n} Re-P_i \times EP-P_i \tag{7.7}$$

where Re-CBS is the reliability of the component-based software, EP-P_i is the probability of the execution of a path and RMC$_i$ is the reusability metric of the component.

7.5.5 Case Studies

To illustrate the proposed method, we use a case study with five components. The component interaction graph (CIG) of the case study is shown in Figure 7.4. To elaborate the role of reusability and interaction issues, we have used all three types of component: new,

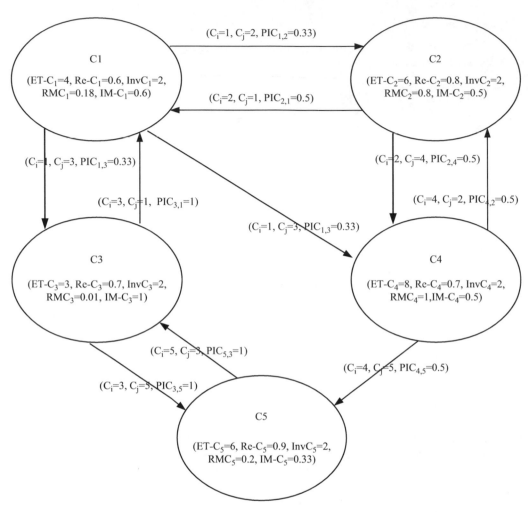

FIGURE 7.4
Component interaction graph.

reused adaptable (both partially qualified and fully qualified) and off the shelf. Component C_3 is the newly developed component, hence its reusability metric is 0.01. Components C_1 and C_5 are partially qualified components having reusability of 0.18 and 0.2 respectively. Component C_2 is fully qualified, with a reusability metric of 0.8, and component C_4 is an off-the-shelf component having reusability metric 1. When one component invokes another component through an out-interaction edge, the calling component will get services in return. We show this type of request through out-interaction edges, and the services in return are shown as in-interaction edges (Figure 7.4). All the interaction edges among components are either in-interaction or out-interaction edges. These interaction edges play an important role in the computation of the interaction metric of these interacting components.

The component interaction graph in Figure 7.4 shows three paths of execution from the starting component C_1 to the ending component C_2.

Based on the execution histories of these paths, we have three cases:

Case 1: $P_1 \rightarrow C_1 \rightarrow C_3 \rightarrow C_5$
Case 2: $P_2 \rightarrow C_1 \rightarrow C_4 \rightarrow C_5$
Case 3: $P_3 \rightarrow C_1 \rightarrow C_2 \rightarrow C_4 \rightarrow C_5$.

Each component, its interaction edge and execution history have corresponding values for these parameters: execution time of C_i, reliability of C_i, number of invocations of C_i for execution and reusability metric of C_i. On the basis of these parameters, we calculate the AT-C_i of component i. These values and computations, which take account of the execution history of the CBS application, are shown in Table 7.10.

CASE 1:   VALUES AND COMPUTATIONS OF PATH P_1

Reliability Computation of Path P_1
First, we calculate the reliability growth (RG_i) in each component due to the reusability metric using Table 7.10 and Table 7.11.
  Reliability growth in component C_1:

$$RGC_1 = (Re-C_i + (1 - Re-C_i^{RMC_i * IM-C_i})) = (0.6 + (1 - 0.6^{0.18*0.6}))$$

$$RGC_1 = 0.65$$

Reliability growth in component C_2:

$$RGC_2 = (Re-C_i + (1 - Re-C_i^{RMC_i * IM-C_i})) = (0.8 + (1 - 0.8^{0.8*0.5}))$$

$$RGC_2 = 0.89$$

TABLE 7.10

Inputs of Path P_1

C_i	ET-C_i	Re-C_i	InvC_i	RMC$_i$	IM-C_i	Av-ET-C_i
C_1	4	0.6	2	0.18	0.6	8
C_3	3	0.7	2	0.01	1	6
C_5	6	0.9	2	0.2	0.33	12

TABLE 7.11

Probability of Component Interaction in Path P_1

$C_i \rightarrow C_j$	PI-C_{ij}
$C_1 \rightarrow C_1$	1
$C_1 \rightarrow C_3$	0.33
$C_3 \rightarrow C_5$	1

Reliability growth in component C_3:

$$RGC_3 = \left(Re- C_i + \left(1 - Re- C_i^{RMC_i *IM-C_i}\right)\right) = \left(0.7 + \left(1 - 0.7^{0.01*0.}\right)\right.$$

$$RGC_3 = 0.7$$

Reliability growth in component C_4:

$$RGC_4 = \left(Re- C_i + \left(1 - Re- C_i^{RMC_i *IM-C_i}\right)\right) = \left(0.7 + \left(1 - 0.7^{1*0.5}\right)\right.$$

$$RGC_4 = 0.86$$

Reliability growth in component C_5:

$$RGC_5 = \left(Re- C_i + \left(1 - Re- C_i^{RMC_i *IM-C_i}\right)\right) = \left(0.9 + \left(1 - 0.9^{0.2*0.33}\right)\right.$$

$$RGC_5 = 0.91$$

Applying these values to Equation (7.6) we calculate the reliability of Path P_1 as

$$EP_1 = \left(0.65 \times 1\right) + \left(0.65 \times 0.7 \times 0.33\right) + \left(0.65 \times 0.7 \times 0.33\right) \times 0.91 \times 1$$

$$EP_1 = 0.94$$

CASE 2:   VALUES AND COMPUTATIONS OF PATH P_2

Applying the values of Tables 7.12 and 7.13 to Equation (7.6), we calculate the reliability of execution history P_2

$$EP_2 = \left(0.65 \times 1\right) + \left(0.65 \times 0.8 \times 0.33\right) + \left(\left(0.65 \times 0.8 \times 0.33\right) \times 0.91 \times 0.5\right)$$

$$EP_2 = 0.90$$

TABLE 7.12

Inputs of Path P_2

C_i	ET-C_i	Re-C_i	InvC_i	RMC$_i$	IM-C_i	Av-ET-C_i
C_1	4	0.6	2	0.18	0.6	8
C_4	8	0.7	2	1	0.5	16
C_5	6	0.9	2	0.2	0.33	12

TABLE 7.13

Probability of Component Interaction in Path P_2

$C_i \rightarrow C_j$	PI-C_{ij}
$C_1 \rightarrow C_1$	1
$C_1 \rightarrow C_4$	0.33
$C_4 \rightarrow C_5$	0.5

Applying the values of Tables 7.14 and 7.15 to Equation (7.6), we calculate the reliability of path P_3

$$EP_3 = (0.65 \times 1) + (0.65 \times 0.8 \times 0.33) + ((0.65 \times 0.8 \times 0.33) \times 0.89 \times 0.5)$$
$$+ ((0.65 \times 0.8 \times 0.33) \times 0.89 \times 0.5) \times 0.91 \times 0.5$$

$$EP_3 = 0.94$$

TABLE 7.14

Inputs of Path P_3

C_i	ET-C_i	Re-C_i	InvC_i	RMC$_i$	IM-C_i	Av-ET-C_i
C_1	4	0.6	2	0.18	0.6	8
C_2	6	0.8	2	0.8	0.5	12
C_4	8	0.7	2	1	0.5	16
C_5	6	0.9	2	0.2	0.33	12

TABLE 7.15

Probability of Component Interaction in Path P_3

$Ci \rightarrow Cj$	PI-Cij
$C_1 \rightarrow C_1$	1
$C_1 \rightarrow C_2$	0.33
$C_2 \rightarrow C_4$	0.5
$C_4 \rightarrow C_5$	0.5

CASE 3: VALUES AND COMPUTATIONS OF PATH P_3

Actual Execution Time and Reliabilities of Paths

Based on the values of the corresponding execution paths, we can compute the execution time, probability of execution, actual execution time and the reliability of each path's execution of the CBS application. These calculations are shown in Table 7.16.

Actual Execution Time and Reliability of CBS Application

Now we can calculate the actual execution time and the reliability of the CBS application using the values of execution histories in Equations (7.5) and (7.7) (Table 7.17).

TABLE 7.16

Execution Time and Reliabilities of Paths

H_i	ET-P_i	EP-P_i	Ac-ET-P_i	Re-P_i
H_1	26	0.33	27.93	0.94
H_2	36	0.33	37.43	0.90
H_3	48	0.33	49.93	0.94

TABLE 7.17

Actual Execution Time and Reliability of CBS Application

Application	Ac-ET-CBS	Re-CBS
CBS	115.29	0.92

Summary

This chapter describes methods for estimating execution time and reliability for component-based applications. The component interaction graph (CIG) not only contains interaction information about components but also holds useful attributes of each coordinating component. Estimates of execution time and reliability are based on the CIG. These computations are based on two fundamental properties of component-based software: reusability and interaction. Reusability metrics and interaction metrics are used to explore the properties of components. The reusability metric of components plays a vital role in the reliability of their execution history and ultimately in the reliability of the CBS application. The results obtained from our study suggest that increased reusability of components supports growth in the reliability of the application because they are pre-tested and qualified components as compared to new components. Since interaction among components through well-defined interfaces contributes to execution time, the interaction metric has been taken into account in the calculation of the time for execution histories and CBS applications.

Although there is a nominal growth in execution time due to these interactions, the interaction metric is only used to assess the interaction of components; it does not restrict their execution. These ratio metrics are not only helpful in computations of the execution and reliability of CBS applications but can be used as an attribute to be stored in the repository along with the component for future reuse.

References

Basili, V. R. and B. Boehm. 2001. "Cots-Based Systems Top 10 List." *IEEE Computer*, 34(5): 91–93.

Chidamber, S. and C. Kemerer. 1994. "A Metrics Suite for Object -Oriented Design," *IEEE Transactions on Software Engineering*, 20(6): 476–493.

D'Souza, D. F. and A. C. Wills. 1998. *Objects, Components, and Frameworks with UML: The Catalysis Approach.* Addison-Wesley, Reading, MA.

Fothi, A., J. Gaizler, and Z. Porkol. 2003. The Structured Complexity of Object-Oriented Programs. *Mathematical and Computer Modeling*, 38: 815–827.

Gill, N.S. and Balkishan. 2008. "Dependency and Interaction Oriented Complexity Metrics of Component-Based Systems." *ACM SIGSOFT Software Engineering Notes*, 33(2): 1–5.

Hamlet, D., D. Mason, and D. Woit. 2001. "Theory of Software Component Reliability." In *Proceedings of 23rd International Conference on Software Engineering*, Toronto, Canada.

Henry, S. and D. Kafura. 1981. "Software Structure Metrics Based on Information Flow." *IEEE Transactions on Software Engineering*, 7: 510–518.

Hopkins, J. 2000. "Component Primer." *Communications of the ACM*, 43(10): 28–30.

IEEE Standard 610.12-1990. 1990. *Glossary of Software Engineering Terminology.* IEEE, New York, ISBN: 1–55937–079–3.

Jacobson, I., M. Griss, and P. Jonsson. 1997. *Software Reuse: Architecture, Process and Organization for Business Success*, ACM Press New York.

Jianguo, C., H. Wang, Y. Zhou, and D. S. Bruda. 2001. "Complexity Metrics for Component-Based Software Systems." *International Journal of Digital Content Technology and Its Applications*, 5(3): 235–244.

Klutke, G., P. C. Kiessler, and M. A. Wortman. 2003. "A Critical Look at the Bathtub Curve." *IEEE Transactions on Reliability*, 52(1): 125–129.

Littlewood, B. and L. Strigini. 1993. "Validation of Ultra-High Dependability for Software-based Systems." *Communications of the ACM*, 36(11): 69–80.

Lyu, M. R. 1996. *Handbook of Software Reliability Engineering.* McGraw-Hill, New York.

Malaiya, Y. K., M. N. Li, J. M. Bieman, and R. Karcich. 2002. "Software Reliability Growth with Test Coverage." *IEEE Transactions on Reliability*, 51(4): 420–426.

McIlroy, M. D. [1968] 1976. "Mass Produced Software Components." In J. M. Buxton, P. Naur, and B. Randell, eds., *Software Engineering Concepts and Techniques*. NATO Conference on Software Engineering, NATO Science Committee, Garmisch, Germany, 88–98.

Meyer, B. 2003. "The Grand Challenge of Trusted Components." In *Proceedings of IEEE ICSE*, Portland, OR, 660–667.

Michael, S. 2000. "Lessons Learned–Through Six Years of Component-Based Development." *Communications of the ACM Journal*, 43(10): 47–53.

Misra, K. B. 1992. *Reliability Analysis and Prediction.* Elsevier, Amsterdam, ISBN: 0–444–89606–6.

Musa, J. 1998. *Software Reliability Engineering.* McGraw-Hill, New York.

Noel, S. 2006. "Complexity Metrics AS Predictors of Maintainability and Integrability of Software Components." *Journal of Arts and Sciences*, 5: 39–50.

Philippe, K. 2003. *The Rational Unified Process: An Introduction*, 3rd ed. Addison-Wesley, Upper Saddle River, NJ.

Sengupta, S. and A. Kanjilal. 2011. "Measuring Complexity of Component Based Architecture: A Graph Based Approach." *ACM SIGSOFT Software Engineering Notes*, 36(1): 1–10.

Tyagi, K. and A. Sharma. 2012. "Reliability of Component Based Systems—A Critical Survey." *WSEAS Transactions on Computers,* 11(2): 45–54.

Vitharana, P. 2003. "Design Retrieval and Assembly in Component-Based Software Development." *Communications of the ACM*, 46(11): 97–102.

Index